Lecture Notes in Engineering

The Springer-Verlag Lecture Notes provide rapid (approximately six months), refereed publication of topical items, longer than ordinary journal articles but shorter and less formal than most monographs and textbooks. They are published in an attractive yet economical format; authors or editors provide manuscripts typed to specifications, ready for photo-reproduction.

Lecture Notes in Engineering

Edited by C. A. Brebbia and S. A. Orszag

59

K. P. Herrmann
Z. S. Olesiak (Eds.)

Thermal Effects in Fracture of Multiphase Materials

Proceedings of the Euromech Colloquium 255
October 31 - November 2, 1989, Paderborn, FRG

Springer-Verlag
Berlin Heidelberg New York London
Paris Tokyo Hong Kong Barcelona

Editors
Klaus P. Herrmann
Laboratory of Technical Mechanics
University of Paderborn
Pohlweg 47-49
4790 Paderborn
FRG

Zbigniew S. Olesiak
Department of Mathematics,
Informatics and Mechanics
University of Warsaw
Palace of Culture and Science
00-901 Warsaw
Poland

ISBN 978-3-540-52858-6   ISBN 978-3-642-88479-5 (eBook)
DOI 10.1007/978-3-642-88479-5

PREFACE

This book contains a selection of the lectures presented at the Euromech Colloquium 255, held at the Liborianum, Paderborn, from 31 October to 2 November 1989. The subject of the Colloquium "Thermal Effects in Fracture of Multiphase Materials" attracted about 50 scientists from 13 countries. Several well known scientists who are active in research on thermal effects in fracture processes were present at the Colloquium as lecturers (29 lectures were delivered) as well as valuable participants of the intensive discussions which took part during the sessions, coffe breaks and lunch times. The closing session of the Colloquium was devoted to a general discussion on the trends in the development of the research in the field, the prospects of the theoretical research, new materials (composites, ceramics etc.), and the trends in technological applications. Over twenty comments and remarks have been made during this final general discussion, showing the interest of the auditorium in such an exchange of viewpoints. However, this discussion is not reflected in this volume.

The Colloquium has been subdivided into six sessions:

I	"Thermodynamics of Fracture Processes"
II	"Fracture of Nonhomogeneous Solids"
III	"Thermal Cracking of Heterogeneous Materials"
IV-VI	"Fracture Phenomena in Composite Systems I-III"

One of the main topics in session I consisted in the description of the influence of thermal effects on shear band localization failure. Thereby shear bands nucleate due to the presence of local inhomogeneities causing enhanced local deformation and local heating where thermal effects influence especially the microdamage process. Further, the localization of plastic deformation within shear bands under the consideration of thermal effects and induced anisotropy on the localisation phenomenon was modelled by a thermoelastoplastic material model for the first stage of the flow process and by a thermoelastic-viscoplastic material model including an advanced micro-damage process for the postcritical behaviour within shear bands.

Moreover, fracture processes in thermoelastoplastic materials were studied by a specialization of the thermodynamic energy balance equations leading to the establishment of a fracture criterion for thermal cracking of nomhomogeneous materials. This criterion also delivers information concerning the prospective crack path of a propagating crack.

Further investigations reported in the sessions II and III concerned fracture mechanisms at free surfaces of heterogeneous materials as well as thermal cracking of solids with local inhomogeneities and microcracks. Thereby discontinuous thermal strains cause microcracking at the surfaces of structures, particularly under cyclic thermal load. By considering a model of a bimaterial interface intersecting a free surface the component of the stress tensor which is parallel to the free surface has a discontinuity at the crossing point, resulting in a very high stress variation under cyclic thermal load which should be the origin for the initiation of edge cracks. Furthermore, discontinuous plastic strains are also the cause of very high stresses at the crossing point where a jump relation between plastic strains and some stress components can be derived which has an analogue in the jump relation for thermal stresses. Special studies dealt with the influence of the cooling rate on the stress intensity factors in thermal shock problems of thwo-phase cracked composite structures. Results were given for a surface crack, a crack terminating at or intersecting the material interface and a crack initiating from the interface.

Besides, the sessions IV-VI of the Colloquium concerned especially thermal cracking of fibrous composites, CFRP-laminates as well as of shape memory alloys and damaged particle strengthened ceramics. Thereby the methods of the micromechanical and the macromechanical approach in composite mechanics, respectively, were applied in order to reach an understanding of the complicated cracking mechanisms under the essential influence of thermal effects. The solutions of the associated boundary value problems of thermoelasticity as well as thermoviscoplasticity were obtained by using modern methods of continuum mechanics, for instance finite element and integral equation methods.

The conference gave an overview about the important influence of thermal effects on the fracture behaviour of multiphase materials. It can be further stated that the scientific and social success of the conference represented a confirmation of the basic idea of Euromech-Colloquia to discuss particular topics in science in small working parties of specialists.

Finally, the two organizers are grateful to the Euromech Committee for approving this Colloquium as well as to the Stiftung Volkswagenwerk, the DAAD, the Nixdorf Computer AG, and the Universitätsgesellschaft Paderborn for some financial support. The help of the administrative staff of the Laboratorium für Technische Mechanik at Paderborn University during the planning phase of the Colloquium and during the Colloquium itself was indispensable.

K. P. Herrmann   Z. Olesiak

CONTENTS

Some Considerations on the Thermodynamics of Fracture

Hein Peter Stüwe *

Abstract

It has become fashionable to discuss mechanical phenomena such as fracture within the framework of thermodynamical equations. This should, of course, never lead to any error; on the other hand, the fertility of such considerations is often overestimated. In this paper it will be tried to illustrate some special points from the experimenter's point of view:

1 Energy necessary to form a fracture surface

The basic term of fracture mechanics is usually written in the form

$$\frac{d\Gamma + dU - dW}{da} \tag{1}$$

where Γ represents the surface energy, U the elastic strain energy and W the potential energy of applied forces. For uniaxial tension on a crack in mode I

$$W = 2U \tag{2}$$

so that (1) reduces to

$$\frac{d\Gamma - dU}{da} \tag{3}$$

(For more complicated situations see, e. g., [1]).

If the term in (1) or (3) is negative, the crack will grow. In the original approach by Griffith, the term Γ was thought to be reversible so that there exists an equilibrium position where (1) or (3) are zero and from which the crack may either shrink or grow leading to

$$\sigma_{Cr} = \sqrt{\frac{2\gamma E}{Aa}} \tag{4}$$

where A is a geometrical factor and γ is the specific surface energy. A typical value for γ is 2 J/m^2; it has been entered in table I.

*Österreichische Akademie der Wissenschaften, Erich-Schmid-Institut für Festkörperphysik, Leoben, Austria

Table 1:

	J/m²
γ	2
γ_B (eq. (5), (6))	2.10^4
U (eq. (8))	2.10^6

In the fracture of real engineering materials γ_B is much higher. It can be determined from a valid K_{IC} test as

$$\gamma_B = \frac{K_{IC}^2}{2E} \tag{5}$$

A K_{IC} test is valid when plastic deformation is limited to a "process zone" near the tip of the advancing crack which is so small that LEFM can be used to anlyse the experiment. Such fractures without large scale plastic flow are often called "brittle" by mechanical engineers. Their fracture surfaces, however, show the dimple structure typical for ductile fracture—at least in materials with high values of K_{IC} as shown is table I.

The energy necessary to form such a fracture surface is [2,3]

$$\gamma_B = \bar{\sigma} h_0 S \tag{6}$$

where $\bar{\sigma}$ is a suitable average flow stress, $S \approx 1/4$ and h_0 characterizes the roughness of the fracture surface. It is therefore possible to determine the K_{IC} value of a fractured material just by studying the geometry of the fracture surface (the stress-strain curve must also be known) [4].

Almost all of this energy is dissipated as heat. Therefore, when γ_B is used in expressions like (1) and (3) it should be kept in mind that it is an irreversible term. (To be exact, a second term should be added in eq.(6) which contains the newly created free surface multiplied by γ. Comparison with table I shows that this second term may be safely neglected!).

2 Softening in low cycle fatigue

Fig. 1(a) shows the result of a test in low cycle fatigue. The apparent flow stress decreases with accumulated strain until the specimen fails catastrophically.

It is tempting to explain such experiments by damage accumulation. Fig. 1(b) shows that it is easy to choose the parameters of damage accumulation theory in such a way that the experiment is well described.

It would be too hasty, however, to take such an agreement as evidence for damage accumulation. This has been shown in the experiment illustrated in fig. 2.

Single crystals of copper of various orientations were predeformed by cold rolling and then subjected to low cycle fatigue. Their flow stress decreased with accumulated strain as shown in fig. 2. The electric resistivity of the samples was measured before and after the test and was seen to decrease during the test. Ascribing the decrease in resistivity to

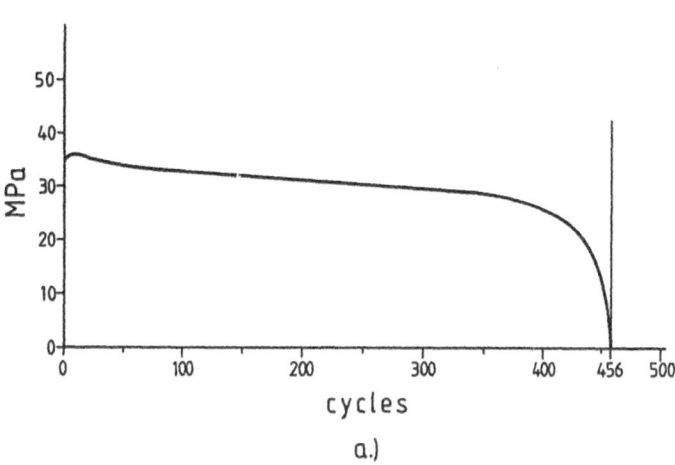

a.)

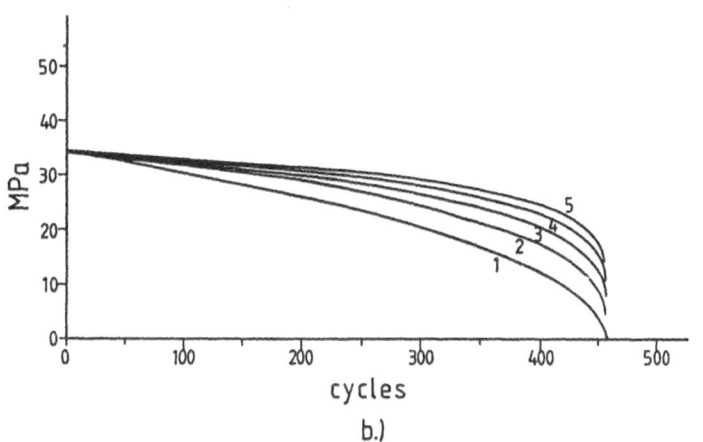

b.)

Figure 1: (a) Critical shear stress of Aluminium subjected to low cycle fatigue in torsion [5] (b) Shear stress vs number of fatigue cycles according to a damage accumulation model [6]

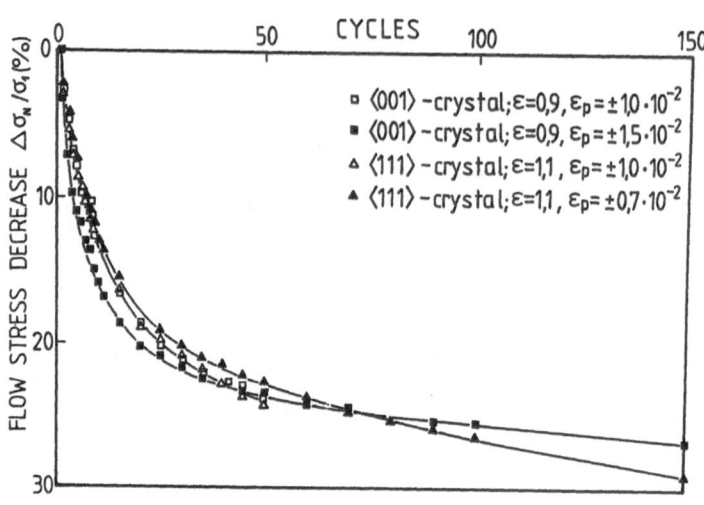

Figure 2: Decrease of flow stress of predeformed copper single crystals during low cycle fatigue[7]

a decrease in dislocation density the change in flow stress could be adequately explained [7].

Thermodynamically to two interpretations are entirely opposite: During damage accumulation the specific internal energy of the material should <u>increase</u> whereas by dislocation annihilation it will <u>decrease</u>.

3   Energy dissipated in fatigue

The energy dissipated during one loading cycle can be quite easily measured in low cycle fatigue as the hysteresis of the stress/strain curve. This becomes increasingly difficult as stresses and strains decrease in high cycle experiments.

Since the plastic strain energy used up in the process zone around the crack tip is almost entirely converted into heat (see chapter 1) the crack front can be considered as a linear heat source. It is then sufficient to observe the temperature field in the specimen to determine the work spent around the crack tip. Several methods to do this have been described in [8].

Fig. 3 shows a typical result. The plastic work spent per cycle and per unit length of crack front $\Delta A_{pl}/B$ is plotted vs. the amplitude of the stress intensity factor ΔK. The data points for the two materials investigated lie very well on two straight lines corresponding to the equation

$$\Delta A_{pl}/B \sim (\Delta K)^4 \tag{7}$$

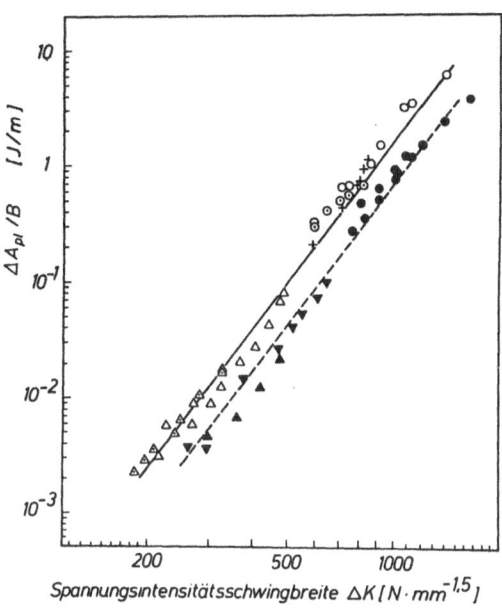

Figure 3: Energy spent at the tip of a fatigue crack in ordinary fatigue[8]. Open symbols: St37, black symbols: St70

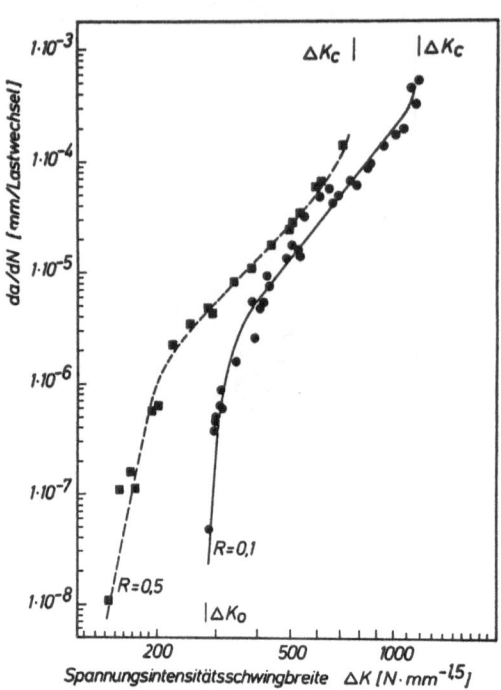

Figure 4: Crack growth rate during fatigue of St70 [8]

Fig. 4 shows the crack growth rates da/dN for the same experiment. The curves have the customary shape; especially they show a central portion where their slope is constant and about equal to 4 (Paris law). Combination with eq.(7) yields the energy spent to create a unit of fatigue fracture surface,

$$U = \frac{\Delta A_{pl}}{B} \frac{dN}{da} \qquad (8)$$

This has been plotted vs. ΔK in fig. 5 [9].

As should be expected U is constant (or nearly so) in the range described by the Paris law. Such a value is therefore shown for comparison in table I. It should not be concluded, however, that this is really the work spent in creating the fracture surface.

Eq.(7) holds even at values of Δk below the threshold value, where the growth rate of the crack drops to virtually zero. This means that a considerable portion of ΔA_{pl} is spent on processes which are geometrically reversible and do not lead to damage even though they do dissipate energy.

(The drop of the curves on the right hand side of fig. 4 is less interesting: it means that

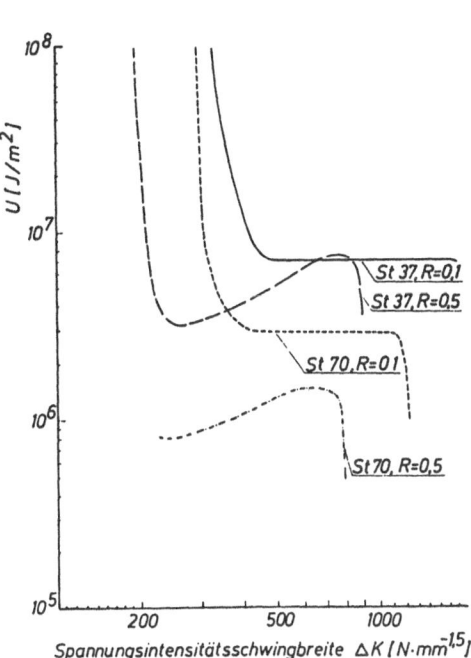

Figure 5: Energy spent during the formation of a unit area of fatigue crack[9]

at high values of stress intensity the fracture mechanism approaches that of unidirectional fracture.)

4 Reversibility of damage

Fig. 6 shows the growth rate of a fatigue crack as a function of crack length or ΔK, respectively. The experiment was performed twice: once in air and once under ultra high vacuum [10]. The two curves are quite similar (they are supposed to show the transition from "short crack behaviour" to "long crack behaviour") but the curve in air is displaced to higher growth rates by a factor of about 5. This means that there are surface reactions affecting crack growth.

Fig. 7 shows schematically how such a surface reaction might work. It shows a crystal with a slip plane before (a) and after glide (b). Upon reversal of stress the crystal slips in the opposite direction. This may happen on the same slip plane (c), which makes the deformation geometrically reversible (although energy is dissipated!). It may also happen on another slip plane leading to intrusions (d) or extrusions (e). Intrusions and extrusions of this type have frequently been observed and are known to contribute to fatigue damage.

The decision whether slip is geometrically reversible (c) or not (d and e) may well depend on the question whether the freshly exposed ledge in (b) is "poisoned" by the adsorption of atoms from the atmosphere or not.

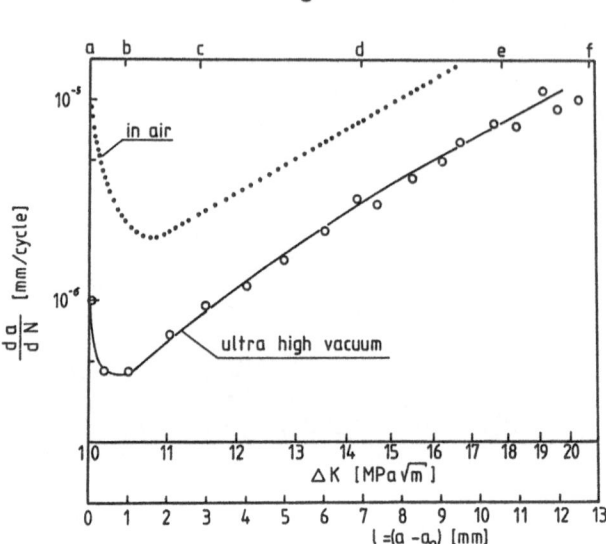

Figure 6: da/dN as a function of crack length at a constant load amplitude test in ultra high vacuum and in air (Armco iron) [10]

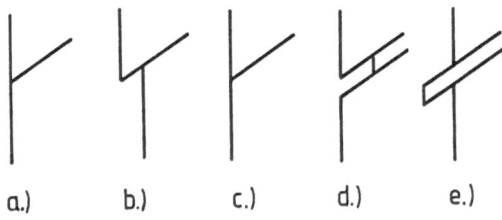

Figure 7: Possibilities for reversal of slip upon reversal of load (see text)

References

[1] W. R. Tyson and G. Roy, Engng. Fracture Mech. 33 (989) p.827

[2] H. P. Stüwe, Engng. Fracture Mech. 13 (1980) p.231

[3] H. P. Stüwe in "Three-Dimensional Constitutive Relations and Ductile Fracture", S. Nemat-Nasser Ed., North Holland Publ. Co. 1981, p.213

[4] O. Kolednik and H. P. Stüwe, Z. f. Metallk. 73 (1982) p.219

[5] K. T. Rie and H. P. Stüwe, Z. f. Metallk. 64 (1973) p.37

[6] F. D. Fischer, personal communication

[7] J. Schrank and H. P. Stüwe, Scripta Met. 23 (1989) p.311

[8] R. Pippan and H. P. Stüwe, Z. f. Metallk. 74 (1983) p.699

[9] R. Pippan and H. P. Stüwe, Berg- und Hüttenm. Monatshefte 129 (1984) p.155

[10] R. Pippan, G. Haas and H. P. Stüwe, Engng. Fracture Mech., in print

INFLUENCE OF THERMAL EFFECTS ON SHEAR BAND LOCALIZATION IN ELASTIC-PLASTIC DAMAGED SOLIDS

Erwin Stein, Maria K. Duszek-Perzyna* and Piotr Perzyna*
Institut für Baumechanik und Numerische Mechanik
Universität Hannover, West Germany

Abstract

The main objective of the paper is the investigation of shear band localization conditions for finite elastic-plastic rate independent deformations of damaged solid body subjected to adiabatic process. For this kind of the process considered the thermal effects may play a dominating role. The objective of the paper is to focus attention on temperature dependent plastic behaviour of a body considered. Thermo-mechanical couplings are investigated and the method is developed which allows to apply the standard bifurcation procedure in examination of the shear band localization criteria when influence of thermo-mechanical couplings and thermal softening effects together with hardening and micro-damage effects are taken into consideration.

Particular attention is focused on the coupling phenomena generated by the internal heat resulting from internal dissipation. A set of the coupled evulation equations for the Kirchhoff stress tensor and for temperature is considered. Assumption that thermo-dynamic process is adiabatic permits to eliminate the rate of temperature and to obtain the general evolution equation for the Kirchhoff stress tensor. The fundamental matrix in this evolution equation describes thermo-mechanical couplings. For the particular elastic properties of the material and for some simplified case of the coupling effects the criteria for shear band localization have been obtained in exact analytical form.

*

On leave from the Institute of Fundamental Technological Research, Polish Academy of Sciences, Warsaw, Poland

1. Introduction

Recent experimental investigations performed under dynamic loading conditions (cf. HARTLEY,DUFFY and HAWLEY [1987], MARCHAND and DUFFY [1988], MARCHAND,CHO and DUFFY [1988]) have shown the importance of thermal effects in the initiation and formation of shear band localization in different steels.

ANAND and SPITZIG [1980] in the investigation of the initiation of localized shear bands in quasi-static plane strain conditions for a maraging steel observed that shear band localization occurs when the hardening modulus rate is decidedly positive, but small and detected small differences between the values of the hardening modulus rate for tension and compression. They also showed that the theoretical predictions from the classical flow theory about the conditions at the initiation of shear bands are not in accord with experiment.

The main objective of the paper is the investigation of the influence of thermal effects on criteria for localization of plastic deformation along shear bands for an elastic-plastic damaged solid body. It is expected that the experimental results concerning the shear band localization phenomenon reported by ANAND and SPITZIG [1980] could be properly explained by considering the influence of two main effects, namely the micro-damage process and thermo-mechanical couplings.

In chapter 2 a thermo-elastic-plastic model of damaged solids is presented. The constitutive equations for elastic-plastic solids, when thermal softening, isotropic and kinematic hardening effects as well as the micro-damage process are taken into consideration, are formulated within a framework of the rate type covariance material structure with internal state variables. The notions of covariance is understood in the sense of invariance under arbitrary spatial diffeomorphism (cf. MARSDEN and HUGHES [1983]). To achieve this aim a multiplicative decomposition of the deformation gradient is adapted and the Lie derivative is used to define all objective rates for intrduced stress and strain measures.

To describe elastic-plastic properties of the material a coupled form of the free energy function is postulated. The isotropic hardening and thermal softening effects are incorporated in the theory directly by defining the hardening-softening material function. The kinematic hardening effect and the softening effect generated by the micro-damage process are described by means of the internal state variable method. The micro-damage process is understood as the nucleation and growth mechanisms of microvoids. Basing on thermodynamic restrictions the general rate type constitutive relation is formulated.

Chapter 3 is devoted to the investigation of thermo-mechanical couplings. The general heat conduction equation is formulated.

In chapter 4 an adiabatic process is investigated. By neglecting the second order terms and concentrating only on main contribution to internal heating the evolution equation for temperature is obtained in a straight-forward form. A set of coupled evolution equation for the Kirchhoff stress tensor and for temperature is considered. The elimination of the rate of temperature leads to the fundamental evolution equation for the Kirchhoff stress tensor. This result allows to use in the examination of the conditions for localization the standard bifurcation method.

In chapter 5 the criteria for shear band localization are investigated. For particular elastic properties of the material and for some simplified case of the coupling effects the criteria for shear bad localization have been obtained in exact analytical form. This result permits to investigate the influence of main cooperative phenomena on the criteria for localization.

Discussions of the results is given in chapter 6. Particular attention is focused on the influence of micro-damage effects and thermo-mechanical couplings.

Chapter 7 brings concluding remarks.

2. Thermo-elastic-plastic model of damaged solids

Let assume that a body is an open set $B \subset \mathbb{R}^3$,and let $\phi : B \to S$ be a C^1 configuration of B in S. The tangent of ϕ is denoted $\mathbf{F}$ and is called the deformation gradient of ϕ .

Let $\{X^A\}$ and $\{x^a\}$ denote coordinate systems on B and S, respectively.

In a neighbourhood of $X \in B$ we consider the local multiplicative decomposition

$$\mathbf{F} = \mathbf{F}^e \cdot \mathbf{F}^p, \tag{2.1}$$

where $(\mathbf{F}^e)^{-1}$ is the deformation gradient that releases elastically the stress in the current configuration $\mathbf{x}$.

Let define the Eulerian strain tensors as follows

$$\mathbf{e} = \frac{1}{2}(\mathbf{g} - \mathbf{b}^{-1}), \quad \mathbf{e}^e = \frac{1}{2}(\mathbf{g} - \mathbf{b}^{e^{-1}}) \quad , \tag{2.2}$$

where $\mathbf{g}$ denotes the metric tensor in the current configuration,$\mathbf{b}$ and $\mathbf{b}^e$ are the total and elastic Finger deformation tensors.

By definition

$$\mathbf{e}^p = \mathbf{e} - \mathbf{e}^e = \frac{1}{2}(\mathbf{b}^{e^{-1}} - \mathbf{b}^{-1}) \tag{2.3}$$

we introduce the plastic Eulerian strain tensor.

The Lie derivative of a spatial tensor field $\mathbf{t}$ with respect to the velocity field $\mathbf{v}$ is defined as [1]

$$L_v\mathbf{t} = \phi_* \frac{d}{dt}\phi^*\mathbf{t} \quad , \tag{2.4}$$

where ϕ^* and ϕ_* denote the pull-back and push-forward operations, respectively.

We have the rates of deformation as follows

$$\mathbf{d} = L_v\mathbf{e} = \frac{1}{2}L_v\mathbf{g}, \quad \mathbf{d}^P = L_v\mathbf{e}^P = \frac{1}{2}L_v(\mathbf{b}^{e^{-1}}), \tag{2.5}$$

and

$$\mathbf{d} = \mathbf{d}^e + \mathbf{d}^p. \tag{2.6}$$

The Lie derivative of the Kirchhoff stress tensor τ gives

$$(L_v\tau)^{ab} = (\overset{\circ}{\tau})^{ab} = [\phi_* \frac{d}{dt}(\phi^*\tau)]^{ab}$$

$$= F^a{}_A F^b{}_B \frac{d}{dt}[(F^{-1})^A{}_c (F^{-1})^B{}_d \tau^{cd}]$$

---

[1] For precise definition of the Lie derivative and its geometrical interpretation please consult ABRAHAM, MARSDEN and RATIU[1988].Application of the Lie derivative to continuum mechanics may be found in MARSDEN and HUGHES [1983] (cf. also SIMO [1988]).

$$= \frac{\partial \tau^{ab}}{\partial t} + \frac{\partial \tau^{ab}}{\partial x^c} v^c - \tau^{cb} \frac{\partial v^a}{\partial x^c} - \tau^{ac} \frac{\partial v^b}{\partial x^c} \quad , \tag{2.7}$$

i.e. the Oldroyd rate (cf. OLDROYD [1950])

To describe inelastic properties of solids it is convenient to introduce a notion of the intrinsic state

$$s = (\mathbf{e}, \mathbf{F}, \vartheta; \boldsymbol{\mu}) \tag{2.8}$$

which consists of as set of variables $(\mathbf{e}, \mathbf{F}, \vartheta; \boldsymbol{\mu})$ given at $\mathbf{x}$ at time t, where $(\mathbf{e}, \mathbf{F}, \vartheta)$ determines the actual deformation-temperature (ϑ denotes absolute temperature) of the considered particle of a body B, and $\boldsymbol{\mu}$ represents a set of the internal state variables, i.e. $\boldsymbol{\mu} \in \mathbf{V}_N$ where $\mathbf{V}_N$ denotes the N-dimensional vector space.

It is postulated that there exists the free energy function

$$\psi = \hat{\psi}(s). \tag{2.9}$$

To specify the evolution equation for the internal state vector $\boldsymbol{\mu}$ we need first its precise physical interpretation.

Let assume

$$\boldsymbol{\mu} = (\boldsymbol{\omega}, \boldsymbol{\alpha}, \xi) \tag{2.10}$$

where the internal state vector $\boldsymbol{\omega} \in V_K$, $K{=}N{-}7$ is introduced to describe the dissipation generated by pure plastic flow phemonena only, $\boldsymbol{\alpha}$ is the residual (or back) stress tensor, and ξ denotes the porosity (the volum fraction).

The internal state vector $\boldsymbol{\mu} \in V_N$ is responsible for all dissipation effects which occur during the thermodynamic process, namely the plastic flow phenomena, the strain induced anisotropy effect and the micro-damage process which consists of the nucleation and growth mechanisms.

Let us introduce the yield criterion in the form as follows

$$\varphi = f(\tilde{\tau}, \mathbf{g}, \vartheta, \xi) - \kappa = 0 \tag{2.11}$$

where

$$f(\cdot) = \tilde{J}_2 + [\, n_1(\vartheta) + n_2(\vartheta) \, \xi \,] \, \tilde{J}_1^2 \quad ,$$

$$\tilde{J}_2 = \frac{1}{2} \, \tilde{\tau}'^{ab} \, \tilde{\tau}'^{cd} \, g_{ac} \, g_{bd}, \quad \tilde{J}_1 = \tilde{\tau}^{ab} \, g_{ab} \, , \quad \tilde{\tau} = \tau - \alpha, \tag{2.12}$$

and κ denotes the work-hardening-softening material function.

The flow rule is postulated in the form

$$\mathbf{d}^p = \Lambda \, \mathbf{P} \, , \tag{2.13}$$

where

$$\mathbf{P} = \frac{1}{2\sqrt{\tilde{J}_2}} \frac{\partial \varphi|_{\xi=const}}{\partial \tau} \, . \tag{2.14}$$

It is assumed that isotropic hardening-softening is described by (cf. NEMES, EFTIS and RANDLES [1988])

$$\kappa = \hat{\kappa}(\varepsilon^P, \xi, \vartheta) = [\kappa_1 + (\kappa_0 - \kappa_1)e^{-h(\vartheta)\varepsilon^P}]^2 \, [1 - \frac{\xi}{\xi^F(\vartheta)}] \, [1 - b\bar{\vartheta}] , \tag{2.15}$$

where κ_0, κ_1 and b are constants, $h = h(\vartheta)$, $\xi^F = \xi^F(\vartheta)$ and

$$\bar{\vartheta} = \frac{\vartheta - \vartheta_0}{\vartheta_0} , \quad \varepsilon^P = \int_0^t (\frac{2}{3} \, d^P : d^P)^{\frac{1}{2}} \, dt. \tag{2.16}$$

The kinematic hardening is postulated in the form of Zigler's rule (cf. ZIEGLER [1959])

$$L_v \, \alpha = a\tilde{\tau} . \tag{2.17}$$

The micro-damage process consists of the nucleation and growth mechanism of microcracks, and is described by the evolution equation for the porosity parameter ξ as follows (cf. GURSON [1975])

$$\dot{\xi} = k_1(s)\tilde{\tau} : d^P + k_2(s)\dot{\tilde{J}}_1 + k_3(s)d^P : g , \tag{2.18}$$

where k_1, k_2 and k_3 denote the material functions. The first term in (2.18) describes debonding of second-phase particles from the matrix material, the second term is responsible for the cracking of second-phase particles and the last term describes the growth process of microcracks.

From the consistency condition

$$\dot{f} - \dot{\kappa} = 0 \tag{2.19}$$

and the geometrical relation (cf. DUSZEK and PERZYNA [1989a])

$$(L_v \, \alpha - rd^P) : Q = 0 , \tag{2.20}$$

where r is new material constant and

$$Q = \frac{1}{2\sqrt{\tilde{J}_2}}[\frac{\partial\varphi}{\partial\tau}|_{\xi=const} + \frac{\partial\varphi}{\partial\xi} \frac{\partial\xi}{\partial\tau}] \tag{2.21}$$

we can determine the coefficients Λ and a.

This leads to the results as follows

$$d^P = \langle \frac{1}{H}\{Q : [\dot{\tau} - (d^e \cdot \alpha + \alpha \cdot d^e)] + \pi\dot{\vartheta}\}\rangle P,$$

$$L_v \, \omega = \Omega(s)\langle \frac{1}{H}\{Q : [\dot{\tau} - (d^e \cdot \alpha + \alpha \cdot d^e)] + \pi\dot{\vartheta}\}\rangle,$$

$$L_v \, \alpha = \frac{H^{**}}{\tilde{\tau} : Q}\tilde{\tau}\langle \frac{1}{H}\{Q : [\dot{\tau} - (d^e \cdot \alpha + \alpha \cdot d^e)] + \pi\dot{\vartheta}\}\rangle, \tag{2.22}$$

$$\dot{\xi} = (k_1\tilde{\tau} : P + k_3 P : g)\langle \frac{1}{H}\{Q : [\dot{\tau} - (d^e \cdot \alpha + \alpha \cdot d^e)] + \pi\dot{\vartheta}\}\rangle + k_2\dot{\tilde{\tau}} : g,$$

where

$$P_{ab} = \frac{1}{2\sqrt{\tilde{J}_2}} \tilde{\tau}^{lcd} g_{ca}\, g_{db} + A\, g_{ab}$$

$$Q_{ab} = \frac{1}{2\sqrt{\tilde{J}_2}} \tilde{\tau}^{lcd} g_{ca}\, g_{db} + B\, g_{ab}$$

$$A = \frac{1}{\sqrt{\tilde{J}_2}}(n_1 + n_2\xi)\tilde{\tau}^{ab} g_{ab}$$

$$B = A + \frac{k_2}{2\sqrt{\tilde{J}_2}}\{n_2\tilde{J}_1^2 + [\kappa_1 + (\kappa_0 - \kappa_1)e^{-h\epsilon^P}]^2 \frac{1 - b\bar{\vartheta}}{\xi^F}\},$$

$$H = H^* + H^{**} \qquad (2.23)$$

$$H^* = -\frac{1}{2\sqrt{\tilde{J}_2}}\{n_2\tilde{J}_1^2 + [\kappa_1 + (\kappa_0 - \kappa_1)e^{-h\epsilon^P}]^2 \frac{1 - b\bar{\vartheta}}{\xi^F}\}[k_1(\sqrt{\tilde{J}_2} + A\tilde{J}_1) + 3Ak_3]$$

$$+ \frac{h(\kappa_1 - \kappa_0)}{\sqrt{3\tilde{J}_2}}[\kappa_1 + (\kappa_0 - \kappa_1)e^{-h\epsilon^P}](1 - \frac{\xi}{\xi^F})(1 - b\bar{\vartheta})(1 + 6A^2)^{\frac{1}{2}}e^{-h\epsilon^P},$$

$$H^{**} = r\mathbf{P} : \mathbf{Q} + \mathbf{Q} : (\mathbf{P} \cdot \boldsymbol{\alpha} + \boldsymbol{\alpha} \cdot \mathbf{P}),\ \pi = \frac{1}{2\sqrt{\tilde{J}_2}}\frac{\partial \varphi}{\partial \vartheta}.$$

To take account of loading criterion the bracket $\langle\rangle$ which defines the ramp function has been introduced. The material function $\Omega(s)$ remains to be determined (cf. DUSZEK-PERZYNA, PERZYNA and STEIN [1989]).

Assuming conservation of mass, balance of momentum, moment of momentum, energy and the entropy production inequality and taking advantage of the postulate that k_2 vanishes when $\tilde{J}_1 \leq \tilde{J}_1^N$ (it means that cracking of second-phase particles takes place when the treshold value for stress is exceded, cf. DUSZEK and PERZYNA [1988]) we obtain the rate type constitutive equation for the Kirchhoff stress tensor τ in the form

$$L_v\tau = \mathcal{L} \cdot \mathbf{d} - z\dot{\vartheta} \qquad (2.24)$$

where

$$\mathcal{L} = [I + \frac{1}{H}\mathcal{L}^e \cdot \mathbf{PQ} - \frac{1}{H}\mathcal{L}^e \cdot \mathbf{PQ} \cdot (\overset{-1}{\mathcal{L}^e} \cdot \boldsymbol{\alpha} + \boldsymbol{\alpha} \cdot \overset{-1}{\mathcal{L}^e})]^{-1} \cdot [\mathcal{L}^e - \frac{1}{H}\mathcal{L}^e \cdot \mathbf{P}(\mathbf{Q} \cdot \tau + \tau \cdot \mathbf{Q})], \qquad (2.25)$$

$$z = [I + \frac{1}{H}\mathcal{L}^e \cdot \mathbf{PQ} - \frac{1}{H}\mathcal{L}^e \cdot \mathbf{PQ} \cdot (\overset{-1}{\mathcal{L}^e} \cdot \boldsymbol{\alpha} + \boldsymbol{\alpha} \cdot \overset{-1}{\mathcal{L}^e})]^{-1} \cdot [\frac{1}{H}\bar{\pi}\mathcal{L}^e \cdot \mathbf{P} + \mathcal{L}^{th}],$$

and

$$\mathcal{L}^e = \varrho_{Ref}\frac{\partial^2 \hat{\psi}}{\partial e^2},\quad \mathcal{L}^{th} = -\varrho_{Ref}\frac{\partial^2 \hat{\psi}}{\partial e\partial\vartheta},$$

$$\bar{\pi} = \pi - \mathbf{Q} : (\overset{-1}{\mathcal{L}^e} \cdot \mathcal{L}^{th} \cdot \boldsymbol{\alpha} + \boldsymbol{\alpha} \cdot \overset{-1}{\mathcal{L}^e} \cdot \mathcal{L}^{th}),\qquad (2.26)$$

$$\varrho_{Ref} = \varrho_M^0(\mathbf{X})(1 - \xi_0) = \varrho_M(1 - \xi)J(\mathbf{X}, t) = \varrho J(\mathbf{X}, t),$$

if $J(\mathbf{X}, t)$ denotes the Jacobian, ϱ_M is the mass density of the matrix material, ξ_0 the initial porosit and ϱ the mass density in the current configuration.

3. Thermo-mechanical couplings

Using the energy balance equation and postulating the Fourier constitutive law for the heat flux we obtain the heat conduction eqution in the form as follows (cf. PERZYNA [1988])

$$\varrho c_p \dot{\vartheta} = div(k\ grad\vartheta) + \vartheta \frac{\varrho}{\varrho_{Ref}} \frac{\partial \tau}{\partial \vartheta} : \mathbf{d}$$

$$+ \varrho \bar{\zeta}_1 \left\langle \frac{1}{H} \{ \mathbf{Q} : [\dot{\tau} - (\overset{-1}{\mathcal{L}^e} \cdot L_v \tau \cdot \alpha + \alpha \cdot \overset{-1}{\mathcal{L}^e} \cdot L_v \tau)] + \bar{\pi}\dot{\vartheta} \} \right\rangle \qquad (3.1)$$

$$+ \varrho \bar{\zeta}_2 [L_v \tau : \mathbf{g} + (\mathbf{g} \cdot \tilde{\tau} + \tilde{\tau} \cdot \mathbf{g}) : \mathbf{d}] \quad,$$

where k is the conductivity coefficient, $c_p = -\vartheta \frac{\partial^2 \hat{\psi}}{\partial \vartheta^2}$ denotes the specific heat and

$$\bar{\zeta}_1 = -\left\{ \left(\frac{\partial \hat{\psi}}{\partial \omega} - \vartheta \frac{\partial^2 \hat{\psi}}{\partial \vartheta \partial \omega} \right) \Omega(s) + \left(\frac{\partial \hat{\psi}}{\partial \xi} - \vartheta \frac{\partial^2 \hat{\psi}}{\partial \vartheta \partial \xi} \right) (k_1 \tilde{\tau} : \mathbf{P} + k_3 \mathbf{g} : \mathbf{P}) \right.$$

$$+ \frac{H^{**}}{\tilde{\tau} : \mathbf{Q}} \left[\frac{\partial \hat{\psi}}{\partial \omega} - \vartheta \frac{\partial^2 \hat{\psi}}{\partial \vartheta \partial \alpha} \right) : \tilde{\tau} - \left(\frac{\partial \hat{\psi}}{\partial \xi} - \vartheta \frac{\partial^2 \hat{\psi}}{\partial \vartheta \partial \xi} \right) k_2 \tilde{\tau} : \mathbf{g} \right] \right\} \quad, \qquad (3.2)$$

$$\bar{\zeta}_2 = -\left(\frac{\partial \hat{\psi}}{\partial \xi} - \vartheta \frac{\partial^2 \hat{\psi}}{\partial \vartheta \partial \xi} \right) k_2 \quad.$$

4. Adiabatic process

Let assume the process considered is adiabatic. This assumption is satisfied for the initial stage of the plastic flow process when the distribution of plastic deformation as well as rate of plastic deformation is homogeneous. Thus the term $div(k\ grad\vartheta)$ in the heat conduction equation (3.1) can be neglected.

For practical application it is resonable to neglect also the second order terms in the heat conduction equation (3.1). Then we concentrate only on main contribution to internal heating which is generated by the internal dissipation.

This leads to the evolution equation for temperature in the form

$$\dot{\vartheta} = \mathbf{M} : L_v \tau + \mathbf{N} : \mathbf{d} \qquad (4.1)$$

where

$$\mathbf{M} = \frac{\zeta_1(\tilde{\tau} : \mathbf{P})\mathbf{Q} + \zeta_2 H \mathbf{g}}{H c_p - \zeta_1 \bar{\pi}(\tilde{\tau} : \mathbf{P})} \quad,$$

$$\qquad (4.2)$$

$$\mathbf{N} = \frac{\zeta_1(\tilde{\tau} : \mathbf{P})(\mathbf{Q} \cdot \tau + \tau \cdot \mathbf{Q}) + \zeta_2 H(\mathbf{g} \cdot \tilde{\tau} + \tilde{\tau} \cdot \mathbf{g})}{H c_p - \zeta_1 \bar{\pi}(\tilde{\tau} : \mathbf{P})} \quad,$$

and

$$\zeta_1 = -\left[\frac{\partial\hat{\psi}}{\partial\omega}\cdot\Omega(s) + \frac{\partial\hat{\psi}}{\partial\xi}(k_1\tilde{\tau}:\mathbf{P}+k_3\mathbf{g}:\mathbf{P})\right.$$
$$\left. + \frac{H^{**}}{\tilde{\tau}:\mathbf{Q}}\left(\frac{\partial\hat{\psi}}{\partial\alpha}:\tilde{\tau} - \frac{\partial\hat{\psi}}{\partial\xi}k_2\tilde{\tau}:\mathbf{g}\right)\right]\frac{1}{\tilde{\tau}:\mathbf{P}} \qquad (4.3)$$
$$\zeta_2 = -\frac{\partial\hat{\psi}}{\partial\xi}k_2 \quad .$$

Substituting (4.1) into (2.24) gives

$$L_v\tau = \mathbf{L}\cdot\mathbf{d} \qquad (4.4)$$

where

$$\mathbf{L} = \left(I - \frac{\mathcal{L}\mathbf{M}}{1+\mathcal{L}:\mathbf{M}}\right)\cdot(\mathcal{L}-\mathcal{L}\mathbf{N}) \quad . \qquad (4.5)$$

It is noteworthy that the evolution equation (4.4) describes for adiabatic process main thermo-mechanical coupling phenomena.

5. Localization criteria

The fundamental result obtained in the form of the evolution equation (4.4) allows to use the standard bifurcation method for the investiagtion of criteria for localization along shear bands in elastic–plastic damaged solids when thermo–mechanical coupling is taken into consideration.

The theory of the localization of plastic deformation along shear bands was developed mainly by RICE [1976], RUDNICKI and RICE [1975], NEEDLEMAN and RICE [1978], RICE and RUDNICKI [1980] and LIPPMANN [1986].

The standard bifurcation method was used by DUSZEK and PERZYNA [1988a, 1988b] to investigate the shear band localization conditions for finite elastic–plastic rate independent deformations of damaged solids. The investigation of the influence of thermo–mechanical coupling effects on the localization criteria for J_2–theory of plasticity was given by DUSZEK and PERZYNA [1989] and for elastic–plastic damaged solids by DUSZEK – PERZYNA, PERZYNA and STEIN [1989].

Let $\mathbf{n}$ be the unit to the surface of a shear band accross which certain components of the velocity gradient may admit jumps but remain uniform outside and inside the band.

Let introduce rectangular Cartesian coordinate system $\{x^i\}$ in such a way that $\mathbf{n}$ is in the x_2–direction. Any non–uniformities in the velocity gradient are kinematically restricted to the form

$$[[\frac{\partial v^i}{\partial x^j}]] = q^i(x_2)\delta_{j2} \quad , \qquad (5.1)$$

where $[[.]]$ denotes the jumps of the enclosed quantity across the discontinuity surface and $q^i(x_2)$ is the magnitude of jump.

The equilibirium condition requires the continuity of the stress rate across the discontinuity surface, i.e.

$$[[\dot{\tau}^{2j}]] = 0 \quad .$$ (5.2)

The conditons (5.1) and (5.2) and the constitutive relation (4.4) give

$$(L^{2jk2} + \tau^{22}\delta^{jk} + \tau^{2j}\delta^{2k})q^k = 0 \quad .$$ (5.3)

The onset of localization occurs at the first instant in the deformation history for which a nontrivial solution of (5.3) exists. Thus, the necessary condition for a localized shear band to be formed is

$$det\left[L^{2jk2} + \tau^{22}\delta^{jk} + \tau^{2j}\delta^{2k}\right] = 0 \quad .$$ (5.4)

To make possible analytical examination of the influence of thermal effects on criteria for localization we have to superpose simplifiations as follows:

(i) Let assume in Eq. (4.1) $\zeta_2 = 0$, it means that in determination of the rate of temperature we do not take into consideration the mechanism of cracking of second phase particles (i.e. postulating $k_2 = 0$).

(ii) Similarly as in the infinitesimal theory of elasticity we postulate

$$(\mathcal{L}^e)^{abcd} = G(g^{ca}g^{db} + g^{cb}g^{da}) + (K - \frac{2}{3}G)g^{ab}g^{cd} \quad ,$$ (5.5)

where G and K denote the shear and bulk modulus, respectively.

(iii) Assume that

$$\overset{-1}{\mathcal{L}^e} \cdot \mathcal{L}^{th} = \Theta g$$ (5.6)

where Θ is the thermal expansion in elastic range.

(iv) Postulate that the Lie derivative is approximated by the material derivative, i.e.

$$L_v\tau \approx \dot{\tau} \quad , \quad L_v\mu \approx \dot{\mu} \quad .$$ (5.7)

Superposition of the simplifications (i) – (iv) leads to the rate equation

$$\dot{\tau} = L^{\#} \cdot d \quad ,$$ (5.8)

where the fundamental matrix $L^{\#}$ is now as follows

$$L^{\#\,abcd} = G(g^{ac}g^{bd} + g^{ad}g^{bc}) + (K - \frac{2}{3}G)g^{ab}g^{cd}$$ (5.9)

$$- \frac{1}{H + G + 9KAB - G\Pi + 6GB\Xi}\left[\frac{G}{\sqrt{\tilde{J}_2}}\tilde{\tau}^{lab} + (3KA + 2G\Xi)g^{ab}\right]\left(\frac{G}{\sqrt{\tilde{J}_2}}\tilde{\tau}^{lcd} + 3KBg^{cd}\right) \quad ,$$

and

$$\Pi = \frac{\zeta_1\pi(\sqrt{\tilde{J}_2} + A\tilde{J}_1)}{c_pG} \quad , \quad \Xi = \frac{3\Theta K\zeta_1(\sqrt{\tilde{J}_2} + A\tilde{J}_1)}{2c_pG}$$ (5.10)

denote the thermal plastic softening and thermal expansion coefficient, respectively.

Setting the matrix $\mathbf{L}^{\#}$ into the Cartesian coordinate system $\{x^i\}$ and substituting into (5.3) we obtain the necessary condition for localization in the form

$$H = \frac{(G\tilde{\tau}'_{22} + 3KA\sqrt{\tilde{J}_2} + 2G\Xi\sqrt{\tilde{J}_2})(G\tilde{\tau}'_{22} + 3KB\sqrt{\tilde{J}_2})}{(K + \frac{4}{3}G)\tilde{J}_2} + \frac{G}{\tilde{J}_2}(\tilde{\tau}^2_{12} + \tilde{\tau}^2_{23})$$

(5.11)

$$- G - 9KAB + G\Pi - 6GB\Xi \quad .$$

Assuming $n_{II} = 0$, postulating that the loading process is nearly proportional and searching for the orientation of the plane within which the shear band localization first takes place by introducing the condition $\frac{\partial H}{\partial n_I} = 0$, i.e. requiring H to be maximum with respect to n_I we finally obtain the necessary conditions for localization

$$\tan^2\beta = \frac{S - \tilde{\tau}'_{III}}{\tilde{\tau}'_I - S} \quad ,$$

(5.12)

$$\frac{H_{cr}}{G} = -\frac{1+\nu}{2}(T + A + B + \frac{1-2\nu}{1+\nu}\Xi)^2 + \frac{1+\nu}{1-\nu}(A - B + \frac{1-2\nu}{1+\nu}\Xi)^2 + \Pi \quad ,$$

where β denotes the angle between the vector $\mathbf{n}$ and the $\tilde{\tau}'_{III}$ direction, ν is Poisson's ratio,

$$S = -(1 - \nu)\tilde{\tau}'_{II} + (1 + \nu)(A + B)\sqrt{\tilde{J}_2} + (1 - 2\nu)\Xi\sqrt{\tilde{J}_2} \quad , \tag{5.13}$$

and

$$T = \frac{\tilde{\tau}'_{II}}{\sqrt{\tilde{J}_2}} \quad . \tag{5.14}$$

6. Discussion of the results

By assuming the isothermal approximation of the process considered we can investigate the influence of micro–damage process on criteria for localization. Then the necessary conditions for localization along the shear band are as follows (cf. DUSZEK and PERZYNA [1988])

$$\frac{H_{cr}}{G} = -\frac{1+\nu}{2}(T + A + B)^2 + \frac{1+\nu}{1-\nu}(A - B)^2 \tag{6.1}$$

and the direction of localization is determined by Eq. (5.12)$_1$ with

$$S = -(1 + \nu)\tilde{\tau}'_{II} + (1 + \nu)(A + B)\sqrt{\tilde{J}_2} \tag{6.2}$$

The result (6.1) for the critical hardening modulus rate $\frac{H_{cr}}{G}$ as function of the state of stress T may be represented by the parabola II, as it has been plotted in Fig

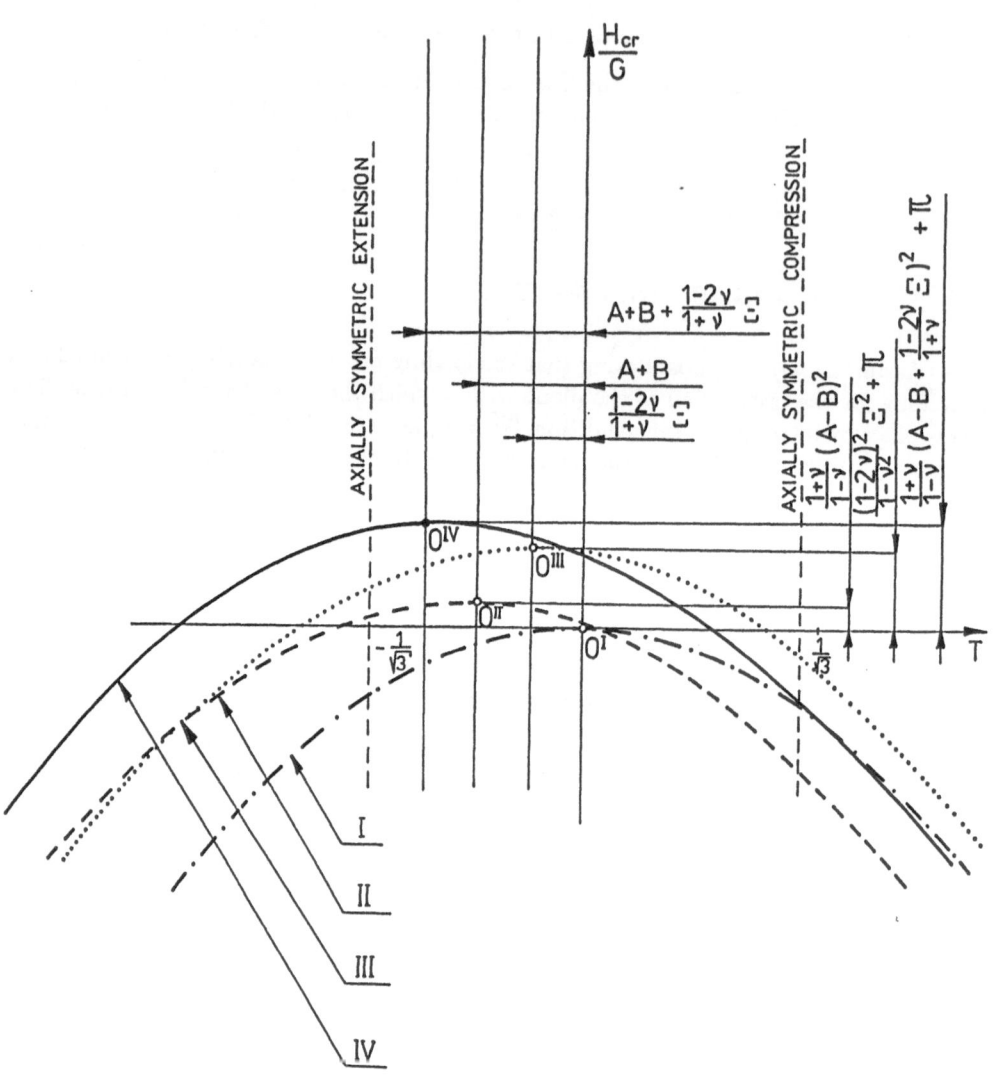

I $\dfrac{H_{cr}}{G} = -\dfrac{1+\nu}{2} T^2$

II $\dfrac{H_{cr}}{G} = -\dfrac{1+\nu}{2} (T+A+B)^2 + \dfrac{1+\nu}{1-\nu} (A-B)^2$

III $\dfrac{H_{cr}}{G} = -\dfrac{1+\nu}{2} (T+\dfrac{1-2\nu}{1+\nu} \Xi)^2 + \dfrac{(1-2\nu)^2}{1-\nu^2} \Xi^2 + \pi$

IV $\dfrac{H_{cr}}{G} = -\dfrac{1+\nu}{2} (T+A+B+\dfrac{1-2\nu}{1+\nu} \Xi)^2 + \dfrac{1+\nu}{1-\nu} (A-B+\dfrac{1-2\nu}{1+\nu} \Xi)^2 + \pi$

Fig.1

1 by dash–dotted line. This parabole when compared with the parabola I given by the equation (for micro–damage process negligible, $A = B = 0$)

$$\frac{H_{cr}}{G} = -\frac{1+\nu}{2}T \qquad (6.3)$$

and plotted in Fig. 1 by broken line, is translated up by $\frac{1+\nu}{1-\nu}(A-B)^2$ and is shifted left by $A + B$.

The translation up is caused by the mechanism of nucleation due to cracking of second phase particles ($k_2 \neq 0$, then $A \neq B$) and shifting left is implied by the micro–damage process (by the nucleation as well as by growth mechanisms).

The translation up means that the material is more inclined to instability by localization along the shear band and the shifting left shows that the inclination to instability for the axially symmetric compression is different from that for the axially symmetric tension. For tension material is more sensitive to localization than for compression. Both these effects have been observed experimentally by ANAND and SPITZIG [1980].

To investigate the influence of thermo–mechanical couplings only on criteria for localization let postulate that there is no micro–damage effects ($A = B = 0$) for adiabatic process. Then the necessary conditions for localization along shear bands are as follows [2]

$$\frac{H_{cr}}{G} = -\frac{1+\nu}{2}\left(T + \frac{1-2\nu}{1+\nu}\Xi\right)^2 + \frac{(1-2\nu)^2}{1-\nu^2}\Xi^2 + \Pi \quad , \qquad (6.4)$$

and the direction of localization is again given by Eq: $(5.12)_1$ with

$$S = -(1-\nu)\tilde{\tau}'_{II} + (1-2\nu)\Xi\sqrt{\tilde{J}_2} \quad . \qquad (6.5)$$

The result (6.4) is plotted in Fig. 1 as the parabola III by dotted line. This parabola, when compared with the parabola I, is translated up by $\frac{(1-2\nu)^2}{1-\nu^2}\Xi^2 + \Pi$ and is shifted left by $\frac{1-2\nu}{1+\nu}\Xi$.

The translation up is caused by both thermal effects, i.e. by thermal expansion (represented by Ξ) and thermal plastic softening (represented by Π), while the shifting left is implied by thermal expansion only.

As it has been already pointed out by DUSZEK and PERZYNA [1989] the experimental results concerning the shear band localization phenomenon reported by ANAND and SPITZIG [1980] can be properly explained by considering the influence of thermo–mechanical couplings only. Indeed, considering the influence of thermal expansion and thermal plastic softening we can have shear band localization when the hardening modulus rate is positive and we have also differences between the values of the hardening modulus rate for axially symmetric compression and tension.

The results obtained in chapter 5 took into consideration micro–damage process and thermo–mechanical couplings simultaneously. Equation $(5.12)_2$ for the critical hardening modulus rate $\frac{H_{cr}}{G}$ as function of the state of stress T can be represented by the

---

[2] These conditions have been first obtained by DUSZEK and PERZYNA [1989].

parabola, cf. the parabola IV plotted in Fig. 1 by solid line. This parabola, when compared with parabola I, is translated up by $\frac{1+\nu}{1-\nu}\left(A-B+\frac{1-2\nu}{1+\nu}\Xi\right)^2 + \Pi$ and is shifted left by $A + B + \frac{1-2\nu}{1+\nu}\Pi$.

For this general case in explanation of the results for the initiation of localized shear band reported by ANAND and SPITZIG [1980] can participate the influence of both effects, namely micro–damage process and thermo–mechanical couplings.

Numerical estimation of the effects which influence criteria for shear band localization has been given by DUSZEK – PERZYNA, PERZYNA and STEIN [1989]. Particular attention has been focused on the comparison of micro–damage and thermo–mechanical coupling effects. In the estimation procedure it has been assumed that $A = B$, hence Eq. $(5.12)_2$ takes the form

$$\frac{H_{cr}}{G} = -\frac{1+\nu}{2}\left(T + 2A + \frac{1-2\nu}{1+\nu}\Xi\right)^2 + \frac{(1-2\nu)}{1-\nu^2}\Xi^2 + \Pi \quad . \tag{6.6}$$

The estimation results are as follows

(i) The micro–damage term $2A$ (cf. Eq. (6.6)) is of the same order as the thermal expansion term $\frac{1-2\nu}{1+\nu}\Xi$.

(ii) The dominated role plays the thermal plastic softening term Π, which at initation of localization is 2.5 times higher than the thermal expansion term $\frac{(1-2\nu)^2}{1-\nu^2}\Xi^2$.

7. Concluding remarks

The investigation of shear band localization conditions for finite elastic-plastic deformations of a damaged solid body subjected to adiabatic thermo-mechanical process has been inspired by the recent experimental results presented by ANAND and SPITZIG [1980], HARTLEY, DUFFY and HAWLEY [1987], MARCHAND and DUFFY [1988] and MARCHAND, CHO and DUFFY [1988].

These experimental results allow us to draw the conclusions as follows:

(i) The predictions from the classical plastic flow theory about the conditions at the initiation of shear band localization are not in accord with experiment.

(ii) In the proper explanation of conditions of shear band localization the dominated role play two effects namely the micro-damage process (softening effects) and thermo-mechanical couplings (thermal expansion and thermal plastic softening, both generated by internal heating effect).

The theoretical results obtained in this paper concerning the influence of micro-damage and thermo-mechanical coupling effects on shear band localization criteria are generally in good agreement with experimental observations.

Particularly, it has been proved that by taking into consideration the micro-damage effects the material is more inclined to instability by localization along the shear band. The localization can occur even for positive value of the rate hardening modulus. It has been also shown that the inclination to instability for the axially symmetric compression is different from that for the axially symmetric tension. For tension material is more sensitive to localization than for compression.

Similarly it has been proved that the experimental results can be properly explained when the influence of thermo-mechanical couplings only is considered. In fact, by taking into consideration the influence of thermal expansion and thermal plastic softening (generated by thermo-mechanical couplings) we can have shear band localization when the hardening modulus rate is positive. On the other hand we have also different results for axially symmetric compression when compared with axially symmetric tension.

The most realistic results have been obtained by taking into consideration the influence of combined micro-damage and thermo-mechanical coupling effects on shear band localization conditions.

Numerical estimation of the effects considered has showed that:

(i) The micro-damage effect is of the same order as the thermal expansion effect.

(ii) The dominated role plays the thermal plastic softening effect, which at initiation of the shear band localization can be 2.5 times higher than the termal expansion effect.

Further numerical estimations and investigations are needed to show quantitative comparison of the influence of micro-damage and thermo-mechanical coupling effects for different state of stress.

The analysis of shear band development in nonhomogeneously deforming solids requires a full initial-boundary value problem solution. Such solution can be obtained only by means of numerical methods.

In recent years the quasi-static as well as dynamic initial-boundary value problems with development of shear bands have been solved by using finite element method, e.g. LE MONDS and NEEDLEMAN [1986], TVERGAARD [1987], NEEDLEMAN [1988,1989] and BATRA and LIU [1989].

Unfortunately the numerical solutions are obtained by superposing some artificial inhomogenities.

To remove this inconvenience one has to develop a method which allows for using in the numerical algorithms the analytical criteria for localization along the shear band.

References

1. R. ABRAHAM, J. E. MARSDEN and T. RATIU, Manifolds, Tensor Analysis and Applications, Springer-Verlag, Berlin 1988.

2. L. ANAND and W. A. SPITZIG, Initiation of localized shear band in plane strain, J. Mech. Phys. Solids, 28 (1980), 113 - 128.

3. R. C. BATRA and DE-SHIN LIU, Adiabatic Shear Banding in Plane strain Problems, J. Appl. Mech., 56 , 1989, 527 - 534.

4. M. DUSZEK and P. PERZYNA, Plasticity of damaged solids and shear band localization, Ing. -Archiv, 58 (1988) 380 - 392.

5. M. DUSZEK and P. PERZYNA, Influence of the kinematic hardening on the plastic flow localization in damaged solids, Arch. Mech., 40 (1988) 595 - 609

6. M. DUSZEK and P. PERZYNA, On combined isotropic and kinematic hardening effects in plastic flow processes, Int. J. Plasticity, (1989) (in print)

7. M. K. DUSZEK and P. PERZYNA, The localization of plastic deformation in thermo-plastic solids, Int. J. Solids and Structures, 1989 (in print)

8. M. DUSZEK-PERZYNA, P. PERZYNA and E.STEIN, Adiabatic Shear Band Localization in Elastic-Plastic Damaged Solids, IBNM Report 89/9; in print Int. J. Platicity, 1989.

9. A. L. GURSON, Plastic flow and fracture behaviour of ductile materials incorporating void nucleation, growth, and interaction; Ph. D. Thesis, Brown University, 1975.

10. K. A. HARTLEY, J. DUFFY and R. H. HAWLEY, Measurement of the temperature profile during shear band formation in steels deforming at high strain rates, J. Mech. Phys. Solids, 35 (1987) 283 - 301.

11. R. HILL, Acceleration waves in solids, J. Mech. Phys. Solids, 10 (1962) 1 - 16.

12. J. LE MONDS and A. NEEDLEMAN, Finite Element Analysis of Shear Localization in Rate and Temperature Dependent Solids, Mech. Math., 56, 1989, 1-9.

13. H. LIPPMANN, Velocity field equations and strain localization, Int. J. Solids Structure, 22 (1986), 1399 - 1409.

14. M. E. MEAR and J. W. HUTCHINSON, Influence of yield surface curvature on flow localization in dillatant plasticity, Mech. Mater. 4 (1985) 395- 407

15. A. MARCHAND, K. CHO and J. DUFFY, The formation of adiabatic shear bands in an AISI 1018 cold-rolled steel, Brown University Report, September 1988.

16. A.MARCHAND and J. DUFFY, An experimental study of the formation process of adiabatic shear bands in a structural steel, J. Mech. Phys. Solids, 36 (1988) 21-283.

17. J. E. MARSDEN and T. J. R. HUGHES, Mathematical Foudnations of Elasticity, Prentice-Hall, N. J. 1983.

18. A NEEDLEMAN and J. R. RICE, Limits to ductility set by plastic flow localization. In Mechanics of sheet metal forming, D. P. Kostinen and N.-M. Wang, eds. pp. 237-267, New York: Plenum 1978.

19. A. NEEDLEMAN, Material Rate Dependence and Mesh Sensitivity in Localization Problems, Comp. Meths. Appl. Mech. Eng., 67, 1988, 69 - 85.

20. A. NEEDLEMAN, Dynamic Shear Band Development in Plane Strain, J. Appl. Mech., 56, 1989, 1 - 9.

21. J. A. NEMES, J. EFTIS and P. W. RANDLES, Viscoplastic constitutiv modeling of high strain-rate deformation, material damage, and spall fracture, 1988 (in print).

22. J. G. OLDROYD,On the formulation of rheological equations of state, Poc. Roy. Soc. (London), Ser. A 200 (1950), 523 - 541

23. P. PERZYNA, Stability of flow processes for dissapative solids with internal imperfections, ZAMP, 35 (1984) 848 - 867

24. P. PERZYNA, Constitutive modelling of dissapative solids for postcritical behaviour and fracture, ASME J. Eng. Mater. Technol., 106 (1984) 410 -419

25. P. PERZYNA, Temperature and rate dependent theory of plasticity of polycrystalline solids, 1989 (in print)

26. J. R. RICE, The localization of plastic deformation, In Theoratical and Applied Mechanics, W. T. Koiter, ed. pp. 207 -220, Amsterdam: North-Holland 1976

27. J. W. RUDNICKI and J. R. RICE, Conditions for the localization of deformation in pressure-sensitive dilatant materials, J. Mech. Phys. Solids, 23 (1975) 371 -394

28. J. R. RICE and J. W. RUDNICKI, A note on some features of the theory of localization of deformation, Int. J. Solids Structures, 16 (1980) 597 - 605

29. J. C. SIMO, A framework for finite strain elastoplasticity based on maximum plastic dissipation and the multiplicative decomposition: Part I, Continuum formulation, Comp. Meths. Appl. Mech. Eng., 66 (1988) 199 - 219

30. V. TVERGAARD, Effect of yield surface curvature and void nucleation on plastic flow localization, J. Mech. Phys. Solids, 35 (1987) 43 - 60

31. H. ZIEGLER, A modification of Prager's hardening rule, Quart. Appl. Math., 17 (1959) 55 - 65

SPECIALIZATION OF THE THERMODYNAMIC ENERGY BALANCE EQUATIONS TO FRACTURE PROCESSES IN THERMOELASTOPLASTIC MATERIALS

B. Kaempf, K.P. Herrmann

Laboratorium für Technische Mechanik, Paderborn University

Pohlweg 47-49, D-4790 Paderborn, F R G

ABSTRACT

Fracture processes are in a special way a consequent effect of the energy distribution around the crack tip. In this respect the strain energy density plays an important role in a cracked thermoelastoplastic solid. In order to formulate a crack driving force the energy balance of the cracked body is used which has to be put in an appropriate form. The latter can be reached by means of the thermodynamic functions like the internal energy as well as the dissipative work. Thereby an extended expression for the dissipative work including irreversible mechanisms at the microscale has been used for the establishment of a fracture criterion for thermal cracking of nonhomogeneous materials. In addition the criterion created delivers informations concerning the prospective crack path of a propagating crack. The theoretical results are illustrated by numerical examples.

1. INTRODUCTION

Manifold fundamental research has been performed including the dissipative effects into the thermodynamic frame. In this context it is worthwhile to study the papers of Lehmann /1/, Müller /2/ and Mazilu /3/ who have done much helpful work in this field. Dui /4/, Curtin /5/ and Mc Cartney /6/ investigated important problems in fracture mechanics. The aim of this paper is to specialize the basic thermodynamic laws to fracture processes in thermoelastoplastic materials. The theory presented here is based on the research report /7/.

Especially in a thermoelastoplastic solid the crack growth is accompanied by plastic deformations which cause dissipative processes. In addition the crack growth can be determined by means of a vectorial fracture criterion which takes into account the dissipative energy. Starting with the general energy balance equation it is possible to formulate and separate the different quantities like the stored energy, the energy flux, the internal energy, etc.. By means of the basic laws of continuum mechanics the special thermodynamic functions will be derived. These functions can be used for the formulation of a global J-integral criterion which is comparable in its special form with the well known J-integral introduced by Rice /8/ and Cherepanov /9/. Further, it is recommendable to study the papers published by Buggisch /10/, De Lorenzi /11/ and Knowles /12/ who did research in the field of the basic theory of the J-integral. In recent years many papers dealing with dissipative fracture processes were published. Besides, the ductile tearing instability /13/ and the J-integral in anisotropic media /14/ have been taken into account, which are important for a fracture assessment in more realistic material models.

2. ENERGY BALANCE FOR FRACTURE PROCESSES

Fracture processes in thermoelastoplastic materials are determined by the dissipated energy and the energy stored in the structure in a special manner. Describing the fracture process by a thermodynamically based theory it is necessary to start with the general energy balance which in its local form reads

$$\rho_0 \dot{\phi} = \psi_{j,j} + \dot{\kappa} \tag{1}$$

with the stored energy ϕ , the energy flux $\psi_{j,j}$ and the energy production of internal sources κ . In eq. (1) the dot means the differentiation with respect to time. For quasistatic processes we can assume that $\dot{\kappa}$ equals zero. The stored energy can be subdivided as follows

$$\dot{\phi} = \dot{\epsilon} + \dot{w}_s + \dot{w}_o + \dot{w}_\beta \tag{2}$$

with the internal energy ϵ , the energy stored in the structure w_s , the surface energy w_o and the energy w_β stored in the structure of the fracture zone. Similarly we receive the formula for the energy flux vector ψ_j

$$\psi_j = \sigma_{ij}\dot{u}_i - q_j \tag{3}$$

with the stress tensor σ_{ij} , the displacement vector u_i and the heat flux vector q_j. It should be mentioned that in eq. (2) the kinetic energy is not taken into account. This can be done easily if necessary. For a further specialization we assume that the internal energy ϵ is a unique function of the elastic strain tensor $\underline{e}$ and the absolute temperature T. The same assumption may hold for the free energy f and the entropy s in a thermoelastoplastic material. The free energy can be subdivided as follows

$$f = f_{e\ell} + f_1(T) \tag{4}$$

with the potential strain energy density $f_{e\ell}$ and the function $f_1(T)$ which will be determined later on.

By introducing the local heat balance the following relation can be derived

$$\dot{\epsilon} = \frac{1}{\rho_o} \sigma_{ij} \dot{e}_{ij} - \frac{1}{\rho_o} q_{j,j} + \dot{w}_D + \dot{w}_\gamma \tag{5}$$

An explicit representation of the thermodynamic functions ϵ, f and s can be given by using Gibb's equation

$$ds = \frac{1}{T}(d\epsilon - \frac{1}{\rho_o} \underline{\sigma}\, d\underline{e}), \tag{6}$$

the thermodynamic relation

$$f = \epsilon - Ts, \tag{7}$$

as well as Hooke's law for thermoelastic materials. After some mathematical operations and by assuming that the potential strain energy $f_{e\ell}$ equals zero during an unrestricted thermal expansion, the functions may be determined by means of the following relations. Thus, by using the approach to the internal energy

$$\frac{\partial \epsilon}{\partial T}\Big|_{\underline{e}} = c_o \tag{8}$$

together with the relation

$$\frac{\partial f}{\partial T} = -s \tag{9}$$

and the requirement for the internal energy

$$\epsilon(o, T_o) = c_o T_o \tag{10}$$

an explicit representation of the functions ϵ, f and s can be obtained

$$\epsilon = w_e + k_o T_o e + c_o T$$

$$f = w_e - k_o e\theta - c_o T \ln\frac{T}{T_o} \tag{11}$$

$$s = k_o e + c_o(1 + \ln\frac{T}{T_o})$$

with the reference temperature T_0, the coefficient of linear thermal expansion α and the relations

$$\theta = T - T_0 \quad ; \quad k_0 = \frac{1}{\rho_0} \frac{E\alpha}{1-2\nu} \quad ; \quad e = e_{pp}$$

$$w_e = \frac{E}{2\rho_0(1+\nu)} \{e_{ij}e_{ij} + \frac{1}{1-2\nu} e^2\} \tag{12}$$

It can be shown that the functions (11) fulfill the second law of thermodynamics. Further, it is interesting to consider the specific heat at constant strain and stress, respectively. Namely, from eq. (11) and the corresponding definition (13)

$$c_\sigma = T \left.\frac{\partial s}{\partial T}\right|_\sigma \quad ; \quad c_e = T \left.\frac{\partial s}{\partial T}\right|_e \tag{13}$$

one obtains

$$c_e = c_0 \tag{14}$$

and

$$c_\sigma = c_0 + 3k_0\alpha T , \tag{15}$$

respectively.

For further considerations it is necessary to determine the potential part $f_{e\ell}$ of the free energy. By means of the requirement that the potential strain energy equals zero during a free thermal expansion $f_{e\ell}$ reads as follows

$$f_{e\ell} = w_e - k_0 e\theta + \frac{3}{2} k_0 \alpha \theta^2 \tag{16}$$

By considering eq. (5) and by combining the irreversible parts outside of the fracture zone

$$\dot{w}_I = \dot{w}_D + \dot{w}_S \tag{17}$$

as well as the corresponding quantities for the specific fracture energy

$$\dot{w}_B = \dot{w}_0 + \dot{w}_\beta + \dot{w}_\gamma \tag{18}$$

the relation for the stored energy (2) leads to the following equation

$$\dot{\phi} = \frac{1}{\rho_0} \sigma_{ij}\dot{e}_{ij} - \frac{1}{\rho_0} q_{j,j} + \dot{w}_I + \dot{w}_B \tag{19}$$

Then, by using the relation for the energy flux (3) from the general energy balance one obtains the following two relations

$$\dot{w}_e - k_0 \theta \dot{e} - \frac{1}{\rho_0} (\sigma_{ij} \dot{u}_i)_{,j} = -\dot{w}_I - \dot{w}_B$$

$$\dot{f}_{e\ell} + k_0 e\dot{\theta} - 3k_0 \alpha \theta \dot{\theta} - \frac{1}{\rho_0} (\sigma_{ij} \dot{u}_i)_{,j} = -\dot{w}_I - \dot{w}_B$$

(20)

3. FORMULATION OF A FRACTURE CRITERION

Now, by changing the differentiation with respect to time to the increment of the increasing crack surface it follows

$$dA_R = 2 \, b \, d \, a$$

(21)

with the width b and the length a of the crack. By introducing a special coordinate system according to (see Fig. 1)

$$x_i = \overset{o}{x}_i - \cos \alpha_i \, \dot{a} \, dt \qquad ; \qquad (i=1,2,3)$$

(22)

the following relation can be derived after some mathematical operations

$$\{f_{e\ell,p} + k_0 e\theta,_p - \frac{3}{2} k_0 \alpha \theta^2,_p - \frac{1}{\rho_0} (\sigma_{ij} u_{i,p})_{,j}\} \cos \alpha_p = w_{I,a} + w_{B,a}$$

(23)

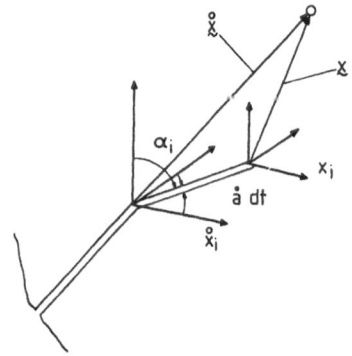

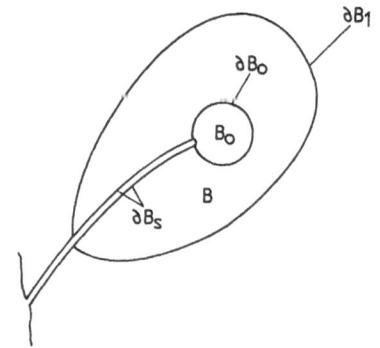

Fig. 1: Reference coordinate system $\overset{o}{x}_i$ and moving system x_i fixed at the moving crack tip.

Fig. 2: Crack tip zone and J-contour integral.
$\partial B^* = \partial B_1 + \partial B_s$

Now the relation (23) has been integrated over the volume B and B_o, respectively, where the corresponding geometry is shown in Fig. 2.

Finally, by using the Green-Gauss theorem the specialized energy balance equation (1) reads

$$L_p \cos \alpha_p = J_p \cos \alpha_p - \int_B \rho_o w_{I,a} dV = 2b\gamma \qquad (24)$$

with the definition

$$J_p = \int_{\partial B*} \{\rho_o [f_{e\ell} - \frac{3}{2} k_o \alpha \theta^2] \delta_{jp} - \sigma_{ij} u_{i,p}\} dA_j + \int_B \rho_o k_o e\theta_{,p} dV \qquad (25)$$

By means of the relation (24) a crack growth criterion can be formulated as follows

$$L_p \cos \alpha_p \geqq 2b\gamma \qquad (26)$$

A comparison with the well-known energy release rate leads to the relation

$$G = \frac{1}{2b} L_p \cos \alpha_p \qquad (27)$$

Further in case of an unstable crack growth the following inequality holds true

$$\frac{dG}{dA_R} > \frac{d\gamma}{dA_R} \qquad (28)$$

Moreover, if the specific fracture energy is independent of the crack length, then it follows from eq. (28)

$$(L_p \cos \alpha_p)_{,A_R} > 0 \qquad (29)$$

4. NUMERICAL EVALUATION OF THE GENERALIZED J-INTEGRAL

In order to study the characteristic behavior of the above mentioned generalized J-integral in elastoplastic materials several FEM-calculations have been performed. Firstly, a crack in a rectangular plate has been investigated (see Fig. 3) by evaluating the J-integral for elastic and elastoplastic material laws, respectively, as well as for displacement and thermal loads. In Fig. 4 two J-curves for elastic and elastoplastic materials, respectively, are shown. It is obvious that the J-values for an elastoplastic material law are higher than for an elastic material law. The difference increases with increasing loads. It must be mentioned here that the J-values were calculated correspondingly to the definition (25). Thereby some authors name J_p the elastic part of the J-integral. In Fig. 5 several J-load-curves corresponding to the specimen shown in Fig. 3 are given where the crack length "a" acts as a parameter. These curves are starting if plastic deformation occurs at the crack tip for the first time. By plotting the J-values versus the crack length a (see Fig. 6) the following conclusions can be stated. For each special load-structure combination there exists a critical crack length which is independent of the applied load.

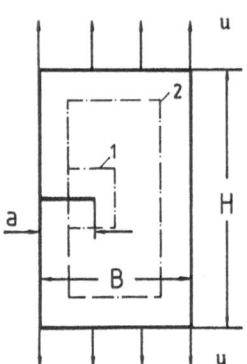

Fig. 3: Edge crack in a rectangular plate

H	=	40.00 mm
B	=	15.00 mm
a	=	5.00 mm
u_m	=	0.01 mm
E	=	70.00 kN/mm^2
y	=	60.00 N/mm^2
	=	0.33

Further calculations have disclosed the interesting phenomenon that the first critical J-value which describes the first occurrence of a plastic deformation is independent of the crack geometry. Because the J-criterion is a vectorial one it is worthwile to verify its applicability by further investigations. Therefore experiments with thermally stressed disklike glass specimens were performed. By using the generalized J-integral (25) a prediction of the prospective crack paths in such glassy compounds was undertaken. Extensive investigations about crack path prediction in two-phase materials have been carried out by Herrmann /15/. Two examples will presented here. The first one is a curved crack path in a disklike specimen consisting of two different optical glasses with a central hole loaded by a homogeneous temperature load. By applying the scalar J-value of the vector-valued J-integral J_p the predicted curved crack path shows a reasonable agreement with the results obtained by the $G_{II} = 0$ crack growth criterion as well as by the experiments, respectively.

The second example is represented by a straight interface crack in a glassy two-phase compound. In this case the calculated J-values show a very good agreement with the numerically obtained values of the total energy release rate G at the crack tip which is under mixed-mode loading.

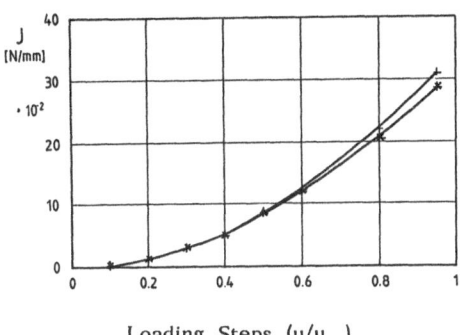

* elastic material law

\+ elastoplastic material law

Fig. 4: J-Integral values in dependence on loading steps

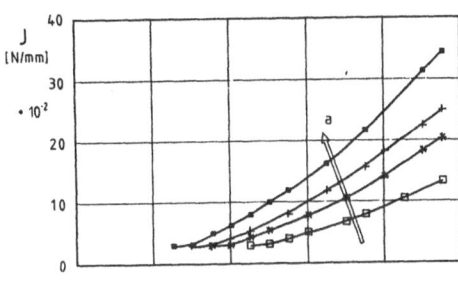

Fig 5a:

Increasing crack length

□ 2 mm; * 3 mm;

\+ 4 mm; ■ 7 mm

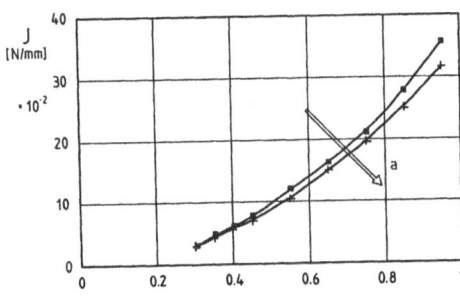

Fig. 5b:

Increasing crack length

\+ 8 mm; ■ 13 mm

Fig. 5a-5b: J-Ingtegral in dependence on loading steps and with the crack length as parameter (Material: Aluminium)

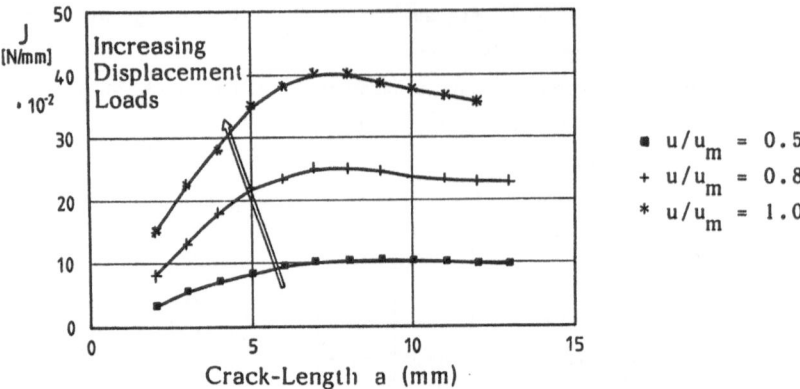

Fig. 6: J-Integral in dependence on crack-length a and with the loading as parameter

CONCLUSION

It has been shown that the formulated generalized J-integral is applicable for a crack path prediction in self-stressed nonhomogeneous solids. Further, the developed theory delivers a crack assessment which takes into account the effect of plastic deformations under thermal as well as mechanical loads. Moreover, the created method allows the calculation of critical crack lengths associated with the specimens chosen. Finally, the energetic basis of the theory is proper for combining the fracture mechanical treatment of a cracked solid with the microstructure where the damage theory could play a crucial role.

ACKNOWLEDGEMENT

The authors gratefully acknowledge the financial help by the Deutsche Forschungsgemeinschaft. Further, they are indebted to M. S. Ming Dong for his support by performing the numerical calculations.

REFERENCES

/1/ Lehmann, Th.: On a Generalized Constitutive Law in Thermoplasticity Taking into Account Different Yield Mechanisms. Acta Mechanica 57, 1-23, 1985

/2/ Müller, I.: "Thermodynamic Theories of Thermoelasticity" in The Constitutive Law in Thermoplasticity. Edited by Th. Lehmann, CISM Course and Lectures No. 281, Springer, 13.104, 1984

/3/ Mazilu, P.: Variationsprinzipe der Plastizität II, Gekoppelte thermodynamische Prozesse. Mitteilungen aus dem Institut für Mechanik, Nr. 37, Universität Bochum, 1983

/4/ Bui, H. D.; Ehrlacher, A.; Nguyen, Q. S.: Propagation de fissure en thermoélasticité dynamique. Journal de Mécanique, 19, 697-723, 1980

/5/ Gurtin, M. E.: Thermodynamics and the Griffith Criterion for Brittle Fracture. Int. J. Solids Structures, 15, 553-560, 1979

/6/ McCartney, L. N.: Discussion:"The Use of the J-Integral in Thermal Stress Crack Problems", by W. K. Wilson and I. W. Yu. International Journal of Fracture 15, R217-R221, 1979

/7/ Herrmann, K.; Kaempf, B.: Theoretische und experimentelle Untersuchungen zum Wärmespannugsbruch von Mehrphasenmedien. DFG-Forschungsbericht, HE 900/7-4, 1988

/8/ Budiansky, B.; Rice, J. R.: Conservation Laws and Energy- Release Rates. J. of Applied Mechanics, 201-203, 1973

/9/ Cherepanov, G. P.: Crack Propagation in Continuous Media. PMM 31, 3, 476-488, 1967

/10/ Buggisch, H.; Gross, D. ; Krüger, K.-H.: Einige Erhaltungssätze der Kontinuums-mechanik vom J-Integral-Typ. Ingenieur-Archiv 50, 103-111, 1981

/11/ DeLorenzi, H. G.: On the Energy Release Rate and the J-integral for 3-D Crack Configurations, Int. J. of Fracture 19, 183-193, 1982

/12/ Knowles, J. K.; Sternberg, Eli: On a Class of Conservation Laws in Linearized and Finite Elastostatics. Archive for Rational Mechanics and Analysis 44, 187-211, 1972

/13/ Zahoor, A.; Paris, P. C.: Ductile Tearing Instability of a Center-Cracked Panel of an Elastic-Plastic Strain Hardening Material: J. of Eng. Mat. and Technology 103, 46-54, 1981

/14/ Tsamasphyros, G.: Path-independent integrals in anisotropic media. Int. J. of Fracture 40, 203-219, 1989

/15/ Herrmann, K.: Thermal Crack Growth in Self-Stressed Glassy Compounds. In: Fracture Mechanics of Non-metallic Materials (Eds. K.P. Herrmann and L.H. Larsson), D. Reidel Publishing Company, Dordrecht, 181-205, 1987

STRESS SINGULARITIES AND DISCONTINUITIES ON THE FREE SURFACE OF A THERMO-ELASTOPLASTIC BIMATERIAL BODY

H.D. Bui[1] and S. Taheri

Electricité de France, Dept. MMN, 92141 Clamart, France

([1]) and Ecole Polytechnique, 91128 Palaiseau, France

Abstract.- Fracture mechanisms at the free surface of heterogeneous materials, poly-cristals, composites etc.., are investigated by consideration of a model of a bimaterial interface intersecting a free surface. The high stress level near the crossing point between the interface and the free surface can be related to discontinuous material properties : elastic moduli, thermal expansion coefficients, plastic behaviors. Analyses of interfaces between two elastic media, which results in logarithmic stress singularities, are widely reported in the literature. This paper is focused on thermal and plastic effects at interfaces. We find that the stress component parallel to the free surface is discontinuous at the crossing point, resulting in a very localized high stress, which we call the "Thorn Singularity". The characteristics of the Thorn singularity are : boundedness of the stress, discontinuity on the free surface, singular gradient of stress. The jump relations between this stress component and the discontinuities of thermal strain and plastic strain are established. Analytical solutions and numerical ones confirm the analyses and show that the stress field can be splitted into a strong local discontinuous field and a global smooth stress field which can be modellized by a classical shell theory.

Introduction

The initiation of cracks is generally observed at the free surface of loaded structures. The microcracks are initiated on defects such as geometrical discontinuities, existing cracks, heterogeneities, inclusions or interfaces of two materials. The interface between dissimilar elastic materials has been widely analysed in the literature, [1] to [5]. There are other interface problems concerning with thermal and plastic behaviors. In this paper, we investigate the stress singularity near the crossing point A, Figure 2a, between the free surface ($x_2=0$) and a surface of discontinuity ($x_1=0$) of thermal strain $[\alpha T]\neq 0$ or plastic strain $[\epsilon^p]\neq 0$, by assuming continuous elastic behaviors. This situation typically corresponds to welding of steel : the basic metal and the weld are generally designed so that they have the same elastic constants, but due to heat and phase transformations the mechanical properties of the weld zone are at lower strength than that of the basic metal, particularly for the yield stress and the hardening coefficient. Very often, the interface between the metal and the weld is the location of crack initiations and growths, Figure 1.

In this paper, we shall examine the stress singularities arising at the interface point A at the free surface. We will find a new type of singularities, weaker than classical unbounded ones. Instead of the logarithmic singularities or negative power behaviors, etc.. we show in this paper that the principal stress parallel to the free surface is continuous inside the body, but discontinuous on the free surface and its gradient is unbounded.

Tube under a thermal shock

In order to study analytically the stress field near the interface z=0 of a discontinuous thermal strain $[\epsilon^T] \neq 0$, we consider a long circular tube (coordinates r, z). The discontinuity of ϵ^T may be caused by either the discontinuity of thermal coefficient $[\alpha] \neq 0$ or the discontinuity of the temperature $[T] \neq 0$ or both $[\alpha T] \neq 0$. Without loss of generality, we consider for example the second case, and we assume that the temperature is prescribed as follows :

$$(1) \qquad T(z) = \frac{\Delta T}{\eta \sqrt{2\pi}} \int_0^z \exp(-u^2/2\eta^2)du$$

where ΔT is the jump $T(+\infty) - T(-\infty)$.

The thermal shock experiment corresponds to the limiting value $\eta \to 0$. Internal and external pressures can be applied to the tube by prescribing $p_i(z)$ at $r=r_i$ and $p_e(z)$ at $r=r_e$. The solution to this thermomechanical problem can be derived in a classical manner, by means of the Love displacement function $\Phi(r,z)$

$$
\begin{aligned}
2\mu\, u_r(r,z) &= - (\partial^2/\partial r \partial z)\Phi(r,z) \\
2\mu\, u_z(r,z) &= 2(1-\nu)\Delta_r\, \Phi(r,z) \\
\Delta_r &= \partial^2/\partial z^2 + \partial^2/\partial r^2 + \partial/r\partial r
\end{aligned}
$$
(2)

The stress components are expressed as follows :

$$
\begin{aligned}
\sigma_{rr}(r,z) &= (\partial/\partial z)(\nu\Delta_r - \partial^2/\partial r^2)\Phi - E\alpha T(z)(1-2\nu) \\
\sigma_{\theta\theta}(r,z) &= (\partial/\partial z)(\nu\Delta_r - \partial/r\partial r)\Phi - E\alpha T(z)(1-2\nu) \\
\sigma_{zz}(r,z) &= (\partial/\partial z)((2-\nu)\Delta_r - \partial^2/\partial z^2)\Phi - E\alpha T(z)(1-2\nu) \\
\sigma_{rz}(r,z) &= (\partial/\partial r)((1-\nu)\Delta_r - \partial^2/\partial z^2)\Phi
\end{aligned}
$$
(3)

The Love function satisfies the equilibrium equation (E : Young modulus, ν : Poisson coefficient)

(4) $\Delta_r \Delta_r (r,z) = E\alpha T'(z)/(1-\nu)(1-2\nu)$

The choice of the particular temperature field (1) is very convenient for the method of solution by Fourier transform $T(z) \rightarrow T^*(s)$

$$T^*(s) = i\Delta T/s\sqrt{2\pi}\, \exp(-s^2\eta^2/2)$$

In the case of thermal shock ($\eta=0$), the analytical solution derived in [6] for the case $p_e = p_i = 0$ shows that ($x_1 \equiv z$, $x_2 \equiv r$) at $x_2 = r_e$ (or $x_2 = r_i$) the Fourier transform of the axial principal stress σ^*_{11} admits asymptotically, for the first term of its Laurent series, the expression

$$\sigma^*_{11}(s,x_2=r_e) \simeq \frac{E\alpha[T]}{1-\nu}\, \frac{i}{2\pi s} + ..$$

This relation in Fourier transform shows that the physical stress $\sigma_{11}(x_1,x_2=r_e)$ is discontinuous at $x_1=0$, $x_2=r_e$, e.g. at the interface point A of the free surface; the discontinuity relation for the general case $[\alpha T] \neq 0$ at the point A is given by

(5) $[\sigma_{11}] = \dfrac{E}{1-\nu}\, [\alpha T]$

The Thorn singularity

In Figure 4. we show the analytical solution $\sigma_{11}(x_1,r_e)$ on the external radius, for $x_1 \geq 0$ (σ_{11} is antisymmetrical with respect to $x_1=0$) compared to the numerical solution by the finite elements method with $\eta=0.1$, with normalized thickness $h/r_m = 0.1$ (r_m : mean radius). For $x_1 \geq \gamma/2$ we recognize the single wave form of the shell solution; near the point $x_1=0$ we observe a very localized and discontinuous stress field $-\sigma_{11}(x_1=-0,r_e) = \sigma_{11}(x_1=+0,r_e) = E[\alpha T]/2(1-\nu)$, see [6]; the characteristic length of the Love-Kirchhoff shell solution is $\gamma = \sqrt{hr_m}$. Along the interface $x_1=0$, due to antisymmetry reason, the stress σ_{11} inside the tube vanishes identically $\sigma_{11}(0,x_2<0)=0$. Fig.3 shows the 3D representation of the field $\sigma_{11}(x_1,x_2)$ inside the tube which indicates the presence of a very localized zone of high stress, which we call suggestively the Thorn Singularity. We will show below that this singularity is characterized by :

-boundedness of the stress
-discontinuity at the free surface point A
-continuity inside the domain
-unbounded gradient.

It is worth noticing that a finite element calculation with a coarse mesh, without the knowledge of the Thorn singularity, would miss totally the local effect. The presence of localized high stress near the interface and the free surface point is a general feature in thermal shock loading, even for arbitrary of geometry of structures, and also in the case of plastic shock as shown in the next section.

Plastic shock

We study the residual stress field in a composite solid, consisting of two materials whose mechanical properties are only different by their plastic behavior. We wish to study the asymptotic behavior of the solution near the point A of the interface intersecting normally the free surface.

We consider first the following problem : Assume that the plastic strain ϵ^P is piecewise constant in each material domain, or more simply $\epsilon^P = \gamma H(x_1)$, where H is the step Heaviside function, H=0 for $x_1 \langle 0$, H=1 for $x_1 \rangle 0$. The domain Ω in consideration is the 3D rectangular bar, Figure 2a. We wish to analyse the residual stress σ_{11} at $x_1 = x_2 = x_3 = 0$. Without loss of generality, we assume that $\gamma = (\gamma_{11}, \gamma_{12}, \gamma_{22}, \gamma_{33})$ is a constant tensor and $\gamma_{ii} = 0$.

The equations are :

$$\sigma = L\epsilon(u) - 2\mu\epsilon^P \text{ in } \Omega$$

(6) $$\text{div } \sigma = 0 \qquad \text{in } \Omega$$

$$\sigma.n = 0 \qquad \text{on } \partial\Omega$$

Let $s = L\epsilon(u)$, $\gamma = \gamma^1 + \gamma^2$, $\gamma^1 = (0, \gamma_{12}, 0, 0)$, $\gamma^2 = (\gamma_{11}, 0, \gamma_{22}, \gamma_{33})$.
We search a solution u^1, for $\epsilon^P = \gamma^1 H(x_1)$, such that :

(7) $$L\epsilon(u^1) = 2\mu\gamma^1 H(x_1) \text{ or } \sigma \equiv 0.$$

Since the right hand side of (7) is a piecewise constant function of x_1, $\epsilon(u^1)$ is a piecewise constant shear strain, thus the above equation is integrable. This means that the plastic deformation $\gamma_1 H(x_1)$ is compatible , in the sense of Kröner. In other words, no residual stress arises from the compatible plastic strain γ^1.

However, there is a residual stress arising from the incompatibility of the plastic strain γ^2. This can be seen by the following physical arguments. Let the solid be separated into two parts along the interface $x_1 = 0$. This operation releases the residual stress and changes the areas of the common interface in different manner, the right bar undergoing the homogeneous deformation γ_{11}, γ_{22}, γ_{33}, Figure 2b. In order to fit together the faces $x_1 = 0$ of separate bars, apply the fictitiuous traction $T_2 = \mp 2\mu\gamma_{22}$ on the faces $n_2 = \pm 1$ of the right half bar , and the fictitious traction $T_3 = \mp 2\mu\gamma_{33}$ on the faces $n_3 = \pm 1$ of the same bar (no traction on the faces $n_1 = \pm 1$). The right half bar is thus subjected to the deformation $(-\gamma_{11}, -\gamma_{22}, -\gamma_{33})$ which makes the surfaces at $x_1 = 0$ superposable; at this stage, the stress σ_{11} is equal to zero. Then after sticking the faces at the interface, we unload these fictitious tractions by applying now the reverse tractions to the reconstituted bar. Precisely during the last operation, the applied traction $T_2 = 2\mu\gamma_{22}H(x_1)$ on the face $n_1 = 1$ induces a discontinuous Neumann boundary condition , hence discontinuous stress components σ_{22}, σ_{11} at the interface point $A(x_1 = x_2 = x_3 = 0)$. The discontinuous stress T_3 applied at far distance from the point A does not induce any discontinuity at A. Finally, because of the discontinuous applied normal traction σ_{22} at A, from the classical theory of elasticity, we obtain the discontinuity relation for the component σ_{11} at the boundary

(8) $\qquad\qquad [\sigma_{11}(A)] = 2\mu[\epsilon^P_{22}]$

The above relation between the discontinuities of stress σ_{11} and plastic strain ϵ^P_{22} has been derived in the simple example of rectangular bar with a piecewise constant plastic deformation. As a matter of fact, this relation is valid for the general case where the plastic strain is discontinuous at the point A, but not necessary a piecewise constant field. In this case, the plastic strain $\epsilon^P(x)$ can always be decomposed into $\epsilon^P = e^1 + e^2$, with e^1 a piecewise constant field and e^2 a continuous field throughout the solid. The component e^2 does not produce any discontinuity of the stress. The component e^1 leads to the discontinuity relation (8) whose validity is thus proved for the general case. This argument is also valid for complex geometry if we make use of asymptotic arguments and assume that the interface is normal to the free surface at the point A.

Numerical calculations

Figure 5 shows the 2D finite element solution for a bimetallic beam, with the same elastic properties but with different plastic characteristics. The plastic deformation ϵ^P_{22} versus x_1 curve has been shifted vertically to show that the discontinuity relation (8) is satisfied with precision.

We find also that the residual stress $\sigma_{11}(x_1, x_2)$ in the beam presents the same thorn singularity as shown in Figure 3.

Expression of the thorn singularity

In the vicinity of the point $x_1 = x_2 = x_3 = 0$, the residual stress may be asymptotically obtained by considering a semi-infinite body. A classical example of a Neumann discontinuous boundary condition on the frontier of the domain $x_2 \leq 0$ is the problem of Boussinesq-Flammant with a constant pressure on $0 < x_1 < a$, $x_2 = 0$. This solution gives us the stress field at $x_1 = 0$

(9) $\qquad\qquad \sigma_{11} = -\pi/2 - \arctg(x_1/x_2) - x_1x_2/(x_1^2 + x_2^2) + \ldots$

The constant term $\pi/2$ has been added to obtain two opposite thorns as shown on Figure 3. In real structures, unsymmetrical solutions may exist and the singularities may have different strengthes. The thorn singularity (9), as indicated before, is bounded at $x_1 = x_2 = 0$, discontinuous on the surface, and its gradient is unbounded.

Discussions

Let us give some interesting values for steel. For a discontinuous temperature of 100 °C with $\alpha = 1.2 \ 10^{-5}$, E=200000 MPa, $\nu=0.3$, the discontinuity of thermal stress is $[\sigma]$=340 MPa . For a discontinuity of α equal to 10^{-6} (the interface weld-metal case) and a service temperature of 550 C°, the discontinuity of σ_{11} is 160 MPa . For a plastic deformation discontinuity of order .001, the stress discontinuity is about 150 MPa . These values show how important is the localized stress at interface, which we call the thorn singularity. Relations (5) and (8) give us the explanation of some phenomena such as the formation of surfaces microcracks in thermal stripping, as observed particularily in cyclic loading, or the initiation of crack at the weld region , Figure 2.

Decomposition of three-dimensional solution in local and global effects

In the previous sections we noticed that the discontinuity of σ_{11} due to a thermal shock is the same in both plane strain and axisymmetric cases. Moreover, it is easy to show that the values of stresses at $x_1=\pm0$ on the external and internal skins for plane strain is the same as in axisymmetrical case, equal to $E[\alpha T]/2(1-\nu)$. This is a quite general result : In fact it can be shown [7] that, for a thin tube ($r_m/h>15$) under a thermomechanical axial loading, in the case of small perturbation, the 3D solution can be decomposed into a local and a global effect. The local effect is independent of the mean radius and may be approximated by a plane strain solution, while the global one may be approximated by a classical Love-Kirchhoff (LK) linear theory. To show this , a decomposition of functions into odd and even functions of the through thickness variable $x_2=r-r_m$ is used through equations (2), (3), (4). The local effect has an extension of an order of the thickness , and is extremely sensitive to the derivative of the temperature profile T"(z).

Figure 6 shows the axial stress on the external skin, for two temperatures fields, which are globally identical but differ only in a small region just by their derivatives. The comparison has been made between three dimensional solutions 3D1, 3D2 and shell solutions LK1, LK2.

For the case of a pure thermal loading, using the above analysis, a shear effect shell approximation can be made and has been given in [7]. The maximum error in comparison with the 3D solution may reach 70% in the case of a thermal shock. This clearly shows the inability of the shell theory to analyse local effects due to discontinuities.

The above mentioned analysis has been extended to the case of thick tube, with the ratio $r_m/h>4$ and also to the case of other structures such as rings and spheres. Using the notion of local effect, it is qualitatively shown that, generally, a classical linear theory is not valid at a clamped edge, because of the existence of a local effect. Figure 7 shows the independence of the local efect with respect to the curvature, in case of a thermal shock. Numerical results reported in Figure 8 shows the decomposition into local and global effect and the parity observed in these effects for a bimaterial tube under traction. Figure 9 shows comparison between solutions for a bimaterial : 1) a 2D plane strain solution, 2) an axisymmetric solution for a tube, 3) a Love-Kirchhoff solution for the tube. The same local effect is found in solutions 1) and 2). Outside the region of local effect, the same global behavior is observed in solutions 2) and 3).

References

[1] F. Erdogan, V. Biricikoglu, Int. J. Eng. Sci., 11, p.745, 1973.

[2] F. Erdogan, G.D. Gupta, Int. J. Fract. 11, p.13, 1975.

[3] R.I. Zwiers, T.C.T. Ting, R.L. Spilker, J. Appl.Mech. Vol.49, p.561, 1982.

[4] M. Comninou et J. D. Dundurs, J. Appl. Mech., 46, p. 849, 1979.

[5] J. R. Willis, J. Mech. Phys. Solids, 19, 1971, p. 353-368.

[6] H.D. Bui, Choc thermique sur le tube générateur de vapeurs, Rapport interne LMS-3, Ecole Polytechnique, 1981.

[7] S. Taheri, Three dimensional local effect and shell theories, Int. J. Ves. and Piping, 36, p. 225-246, 1989.

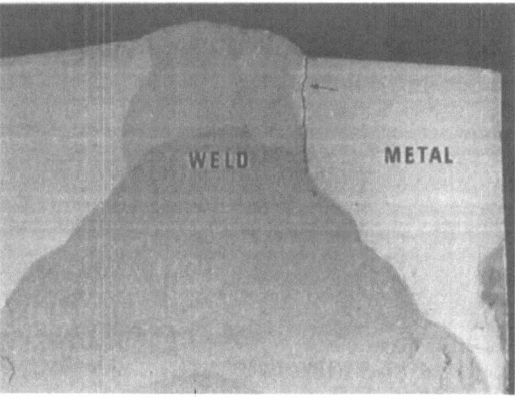

Figure 1 :   Crack initiation at the weld/metal interface.

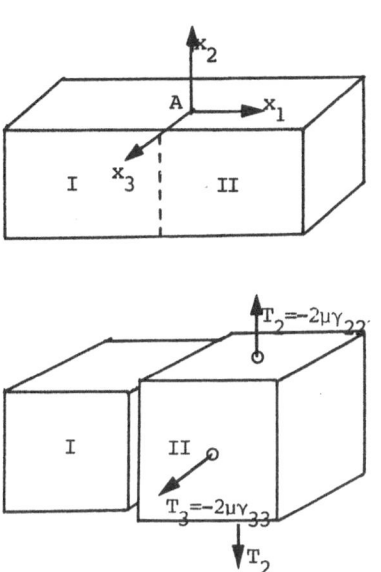

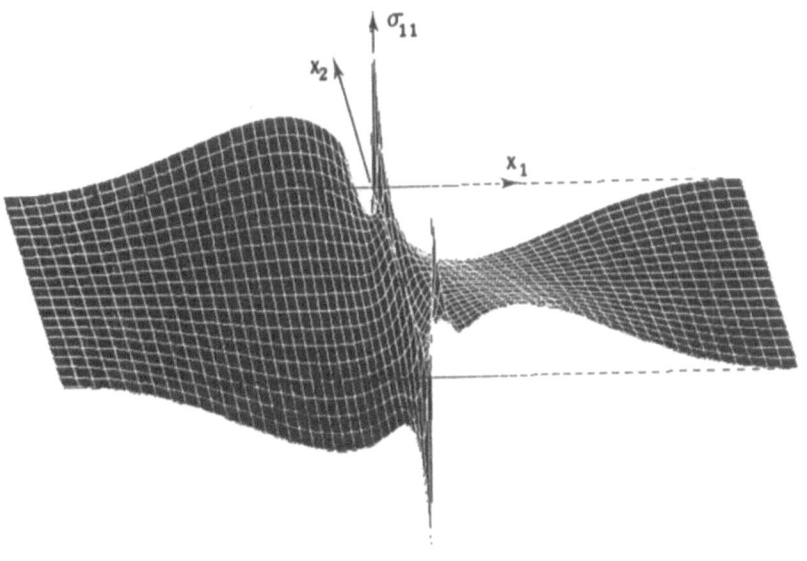

Figure 2 :   a) Interface intersecting a free surface
               b) Release of residual stress, fictituous tractions.

Figure 3 :   The Thorn Singularities of the stress $\sigma_{11}(x_1, x_2)$.

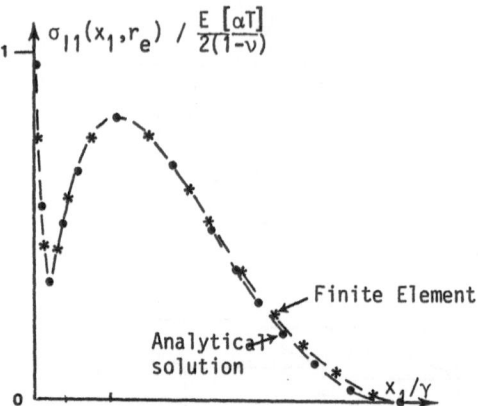

Figure 4 :   Axial stress at external radius of a tube in thermal shock

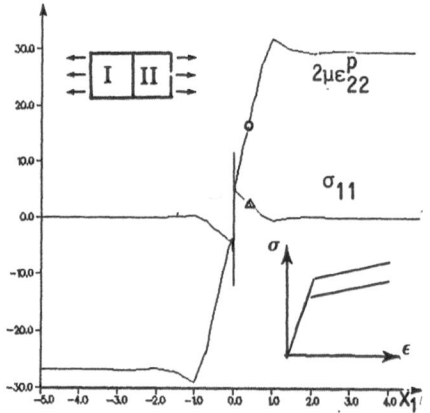

Figure 5 :   Comparison between σ_{11} and $2\mu\epsilon^P_{22}$ at the free surface

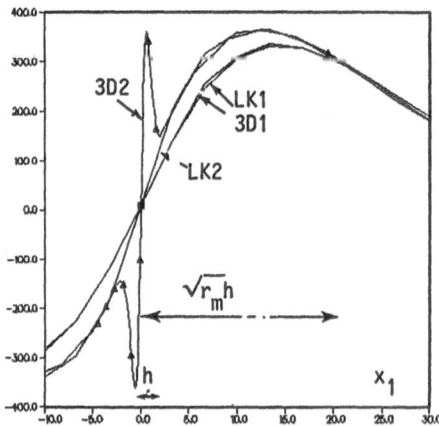

Figure 6 :   Comparison between 3D and LK solutions for 2 temperatures.

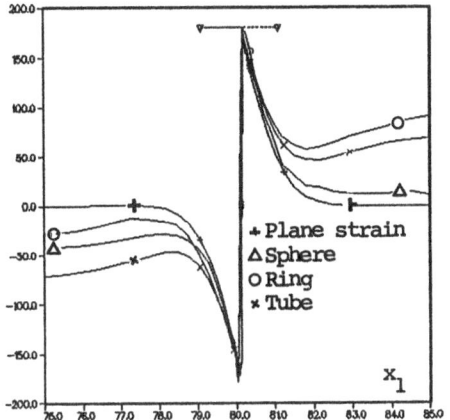

Figure 7 : Curvature-independent local effect.

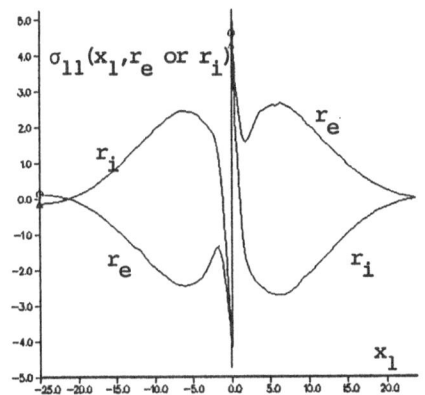

Figure 8 : Symmetric local effect and antisymmetric global effect in tube with respect to the through thickness variable $r-r_m$.

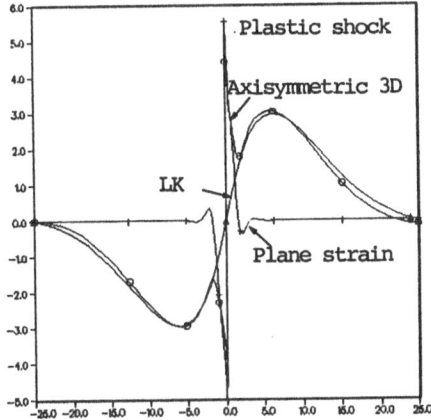

Figure 9 : Comparison of solutions in different theories : i) 2D plane strain ; ii) Axisymmetric 3D solution ; iii) LK solution .

STRESS-DRIVEN SOLUTE SEGREGATION NEAR CRACK-TIPS:
THEORETICAL MODELLING AND APPLICATIONS TO THE INTERGRANULAR
FRACTURE OF HETEROGENEOUS MATERIALS

H. Rauh

Materials Development Division, Harwell Laboratory
Oxfordshire OX11 0RA, England

and

Department of Metallurgy & Science of Materials
University of Oxford, Oxfordshire OX1 3PH, England

Abstract

Recent progress in understanding stress-driven solute segregation during intergranular fracture is reviewed. Theoretical predictions and model assumptions are related to experimental observations of cracking of ferritic steels and discussed in the light of available evidence.

Introduction

It is now widely appreciated that stress-driven solute segregation near crack-tips plays a key role for the mechanical behaviour of materials where performance depends on bulk transport of embrittling agents to fracture zones. Originally suggested and demonstrated with the example of hydrogen in cracked specimens of iron under stress [1,2], this mechanism has received significant attention following the discovery of slow, high-temperature brittle intergranular fracture of low-alloy steels; a novel type of failure which requires local enrichment of impurities such as sulphur in the region of intensified stress around crack-tips [3,4].

The concept underlying such segregation and concomitant grain-boundary fracture events proceeds from the stress field of a loaded crack interacting with nearby point-defects. At temperatures ensuring sufficient mobility, this interaction can impose a drift flow upon the point-defects that determines their migration in the vicinity of the tip. Thus, enhanced segregation to the crack as well as to the grain-boundary ahead of the crack may occur. When the defects are an embrittling solute its accumulation can, in turn, promote fracture by reducing grain-boundary cohesion and encouraging crack propagation under the prevailing conditions of stress intensity and temperature.

In this paper I take the opportunity to highlight briefly some of the recent progress in modelling and application that has been achieved, with emphasis on work at Harwell. The segregation process triggering intergranular fracture, as previously invoked and quantified by us [4,5], will be outlined for both undersized and oversized solute. Theoretical predictions and inherent assumptions will be related to experimental observations of cracking of ferritic steels and discussed in the light of available evidence.

Model

Consider a long, straight, semi-infinite crack within an isotropic elastic body and a coplanar, unbroken grain-boundary ahead of the crack. Under mixed-mode loading by an applied uniaxial tension, intrinsic to the meandering process of intergranular fracture addressed, the crack-tip exerts a stress field characterized by the stress-intensity factor K. When the axis of applied tension is orthogonal to the crack-tip and inclined at an angle α to the grain-boundary half-plane (where $0 < \alpha < \pi/2$, without loss of generality), the energy of interaction between this stress field and a point-defect, represented by a misfitting spherical inclusion located at distance r from the tip and azimuth θ to the boundary, has the harmonic form [5]

$$E_\alpha = A \sin\alpha \, \sin(\theta/2-\alpha)/r^{1/2}; \quad -\pi < \theta < \pi, \tag{1}$$

in which
$$A = (2/9\pi)^{1/2}(1+\nu)K\Delta V, \tag{2}$$

where ν denotes Poisson's ratio of the elastic body and ΔV the relaxation volume of the point-defect.

To analyse the kinetics of segregation, we envisage transient depletion of solute from an initial uniform solute atom concentration. Assuming solute flow near the crack-tip arises predominantly from the crack-tip stresses and random diffusion may be ignored, the solute current density, in the 'pure-drift' approximation, is given by the Einstein equation [6,7]

$$j_\alpha = - (D/k_BT)c_\alpha \nabla E_\alpha \tag{3}$$

with the solute diffusion coefficient D, Boltzmann's constant k_B and absolute temperature T; the volume concentration of solute c_α itself is determined by the continuity equation

$$\partial c_\alpha/\partial t + \nabla \cdot j_\alpha = 0 \tag{4}$$

recalling expression (3), together with the condition $c_\alpha = c_o$ at initial time $t = 0$, introducing a positive constant c_o, and the requirement $c_\alpha = c_o$ as $r \to \infty$ at any time $t > 0$.

The appropriate solution for the transient solute atom concentration c_α around the crack-tip (implying that both the crack and the associated grain-boundary act as ideal point-defect sinks) consists of two regions separated by a characteristic which rigidly expands with time; in the inner region the concentration is identically zero and in the outer region it retains its initial value c_o. Fig. 1 illustrates this solution at a particular time $t > 0$ for $\alpha = \pi/3$ and relaxation volumes of either sign; the figure also shows equipotentials derived from eqn. (1) and flow lines given by orthogonal trajectories to the equipotentials along which solutes drift in the arrowed directions, thereby depleting the region inside the characteristic. Evidently, when $\Delta V < 0$ the solute atoms enter across the crack-surfaces and along the grain-boundary from below, whereas when $\Delta V > 0$ they enter along the grain-boundary from above and into the crack-tip [5,7].

The numbers of solute atoms lost at the various sinks after time $t > 0$, per unit length of the crack-tip in excess of the numbers deposited at time $t = 0$, equal the numbers of such point-defects initially present within the respective sections enclosed by the expanding characteristic. Thus, $N_\alpha^{uc}(t)$, the number of solute atoms segregated to the upper crack-surface (when $\Delta V < 0$) or $N_\alpha^{gb}(t)$, the quantity lost at the grain-boundary from above (when $\Delta V > 0$), in time $t > 0$, is [5]

$$N_\alpha^{uc}(t) | N_\alpha^{gb}(t) = \tfrac{5}{4}(\tfrac{1}{4}f(\alpha))^{1/5} (\sin\alpha)^{4/5} c_o [\tfrac{ADt}{k_BT}]^{4/5}, \qquad (5)$$

where $\qquad\qquad\qquad f(\alpha) = 3\pi + 8\sin(2\alpha). \qquad\qquad\qquad (6)$

Similarly, $N_\alpha^{\ell c}(t)$, the number of solute atoms segregated to the lower crack-surface (when $\Delta V < 0$) or $N_\alpha^{ct}(t)$, the quantity lost at the crack-tip from behind (when $\Delta V > 0$), in time $t > 0$, is [5]

$$N_\alpha^{\ell c}(t) | N_\alpha^{ct}(t) = \tfrac{5}{4}(\tfrac{1}{4}g(\alpha))^{1/5} (\sin\alpha)^{4/5} c_o [\tfrac{ADt}{k_BT}]^{4/5}, \qquad (7)$$

where $\qquad\qquad g(\alpha) = 6\alpha - 5\sin(2\alpha) + 2\sin(2\alpha)\cos^2\alpha. \qquad\qquad (8)$

Likewise, $N_\alpha^{gb}(t)$, the number of solute atoms segregated to the grain-boundary from below (when $\Delta V < 0$) or $N_\alpha^{ct}(t)$, the quantity lost at the crack-tip from the front (when $\Delta V > 0$), in time $t > 0$, is [5]

(a)

(b)

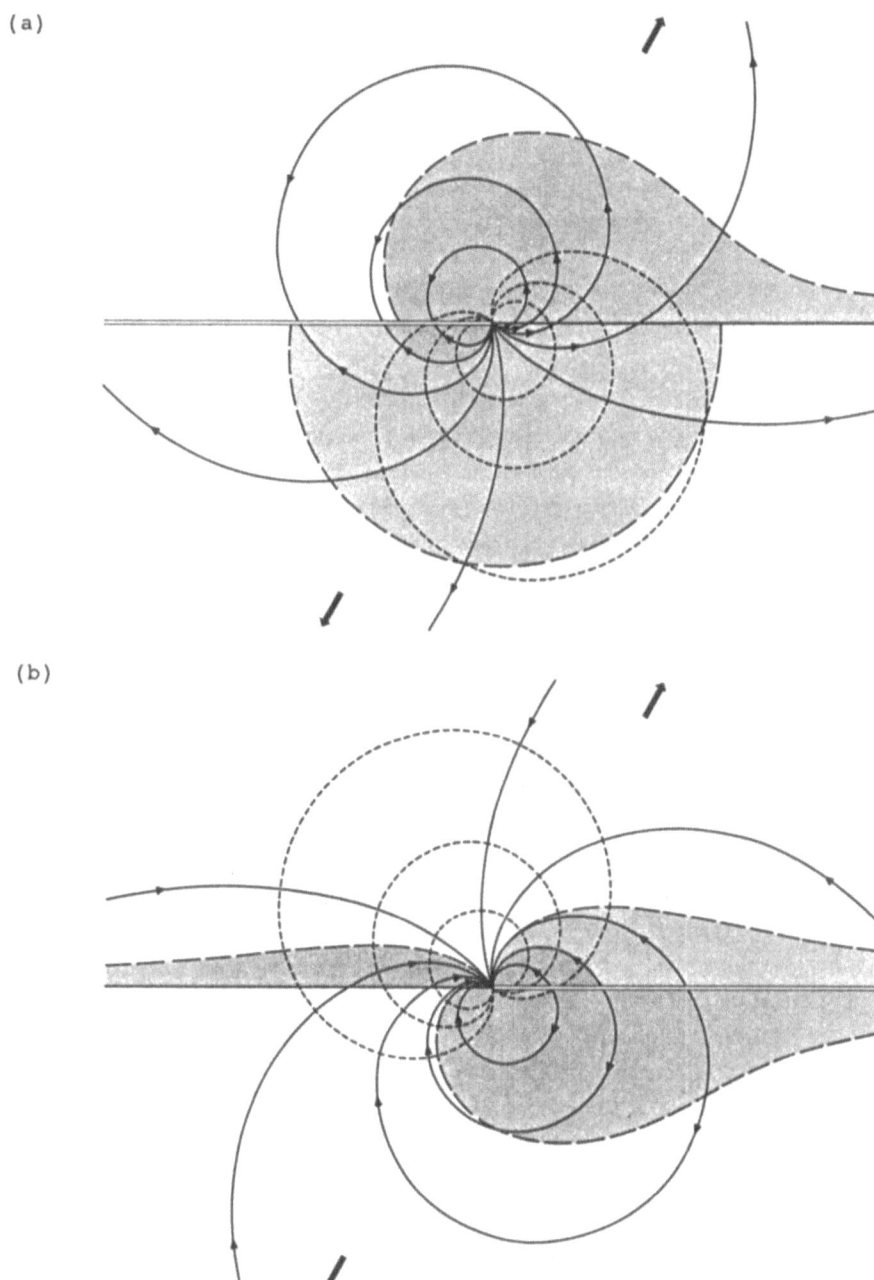

Fig. 1. Equipotentials (----), flow lines (⟶), and expanding characteristic (— —) in the vicinity of the crack (straight double lines) subject to mixed-mode loading by an applied uniaxial tension (bold arrows), and the unbroken grain-boundary (straight single line) ahead of the crack during transient depletion of solute atoms, when their relaxation volume (a) $\Delta V < 0$ and (b) $\Delta V > 0$. The solute concentration is zero inside the characteristic (shaded region) and equal to its initial value c_o outside.

$$N_{\alpha}^{gb}(t)\,|\,N_{\alpha}^{ct}(t) = \frac{5}{4}(\frac{1}{4}h(\alpha))^{1/5}\,(\sin\alpha)^{4/5}\,c_o\,[\frac{ADt}{k_BT}]^{4/5}, \qquad (9)$$

where
$$h(\alpha) = 3(\pi-2\alpha) - 5\sin(2\alpha) + 2\sin(2\alpha)\sin^2\alpha. \qquad (10)$$

$N_{\alpha}^{tot}(t)$, the total number of solute atoms with $\Delta V < 0$ or $\Delta V > 0$ segregated to all respective sinks in time $t > 0$, follows as the sum of the individual contributions, eqns. (5), (7) and (9),

$$N_{\alpha}^{tot}(t) = N_{\alpha}^{uc}(t)\,|\,N_{\alpha}^{gb}(t) + N_{\alpha}^{\ell c}(t)\,|\,N_{\alpha}^{ct}(t) + N_{\alpha}^{gb}(t)\,|\,N_{\alpha}^{ct}(t); \qquad (11)$$

its variation with angle α is displayed in Fig. 2. We note, when $\alpha = 0$ no solute is lost, since the interaction energy (1) then precludes stress-driven segregation to any sink. With increasing α, i.e. mixed-mode loading and hence a driving force present, the total solute atom loss rises up to a maximum at $\alpha = 0.41\pi$, followed by a slight reduction as α further increases towards the mode I loading orientation $\alpha = \pi/2$. The variation with angle α of the numbers of solute atoms segregated to the grain-boundary in time $t > 0$, eqns. (5) and (9), is depicted in Fig. 3. These losses have a maximum at $\alpha = 0.22\pi$ (when $\Delta V < 0$) or $\alpha = 0.42\pi$ (when $\Delta V > 0$) and vanish at $\alpha = 0$, as explained before.

To utilise the above results for the problem of solute segregation during intergranular crack propagation, we envisage crack growth to proceed in a 'step-wise' fashion, with average velocity v, by sudden jumps and stops: once embrittled locally through enrichment with solute at a level sufficient for decohesion of the unbroken grain-boundary in front, the crack jumps forward a discrete distance and arrests, waiting for further segregation during a time Δt, whereupon the process repeats. Provided that each jump takes the crack-tip well into regions of fresh (and hence tougher) material where negligible depletion has occurred and the solute concentration is therefore at c_o, the average segregation can be deduced from the sequence of (small-time) depletions between jumps, when the crack is stationary, using eqn. (11) [4,5].

Two features are hereby significant. On the one hand, since maximum solute segregation to all locations near the crack-tip and to the grain-boundary ahead of the tip occurs when the crack-plane deviates from the mode I loading orientation (cf. Figs. 2 and 3), crack growth will be favoured across boundaries that do not lie parallel to such a plane. The decohesion mechanism associated with the fracture event studied here, on the other hand, requires a preference for the loading component of maximum tensile stress. It is thus clear that most of the grain-boundaries potentially suitable for intergranular crack

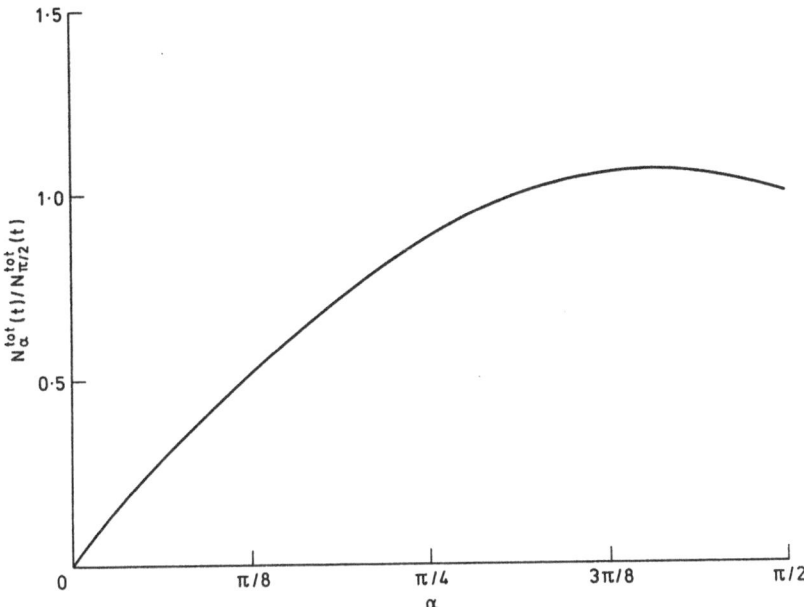

Fig. 2. Total number of solute atoms with $\Delta V < 0$ or $\Delta V > 0$ segregated to all respective sinks in time $t > 0$, as as function of the angle α, normalized with the overall solute atom loss for $\alpha = \pi/2$.

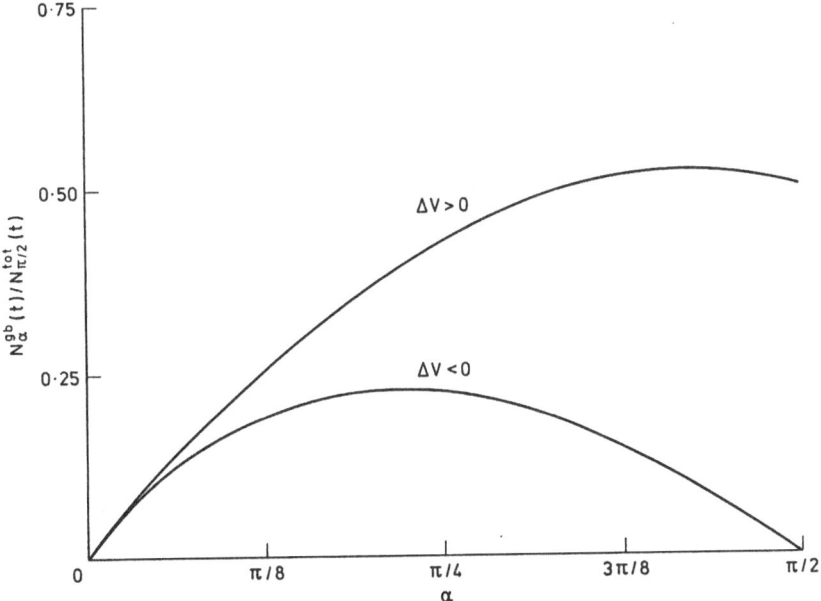

Fig. 3. Numbers of solute atoms segregated to the grain-boundary from below (when $\Delta V < 0$) and from above (when $\Delta V > 0$) in time $t > 0$, as a function of the angle α, normalized with the overall solute atom loss for $\alpha = \pi/2$.

propagation will lie reasonably close to the mode I loading orientation, occupying the range $\pi/4 \lesssim \alpha \leq \pi/2$, say. This has a straight-
forward implication for the theoretical prediction of the segregant
coverage on fracture surfaces; a quantity determined experimentally (if
feasible) as an average over several crack jumps across a single
intergranular facet at fixed, albeit usually unknown, angle α. Since,
for α within the range considered, the total number of solute atoms
segregated to all sinks varies only little from its value at $\alpha = \pi/2$
(cf. Fig. 2), the expected average total solute coverage, per unit area
of crack path, for the mixed-mode loading situation will be

$$\langle P \rangle = \langle N_\alpha^{tot}(\Delta t) \rangle / v \Delta t \approx N_{\pi/2}^{tot}(\Delta t) / v \Delta t, \qquad (12)$$

i.e. essentially the same as if mode I loading prevailed [5].

Applications

We now relate the model predictions and underlying assumptions to
experimental observations of two particular fracture phenomena in
heterogeneous materials which are of both current interest and
practical importance.

a) High-temperature brittle intergranular fracture

This phenomenon, first seen as stress-relief cracking within the
coarse-grained heat-affected zone of welds, is most prominent in 'as-
quenched' microstructures of low-alloy steels subject to stress
concentrators at temperatures between 300 and 650°C [8]; it depends
primarily on the stress-driven segregation of sulphur near crack-tips,
together with segregation of residual impurities (e.g. phosphorus, tin
and antimony) to grain-boundaries under thermal activation alone [4,9].

Consider a notched bend specimen of CrMo steel austenitised at
1200°C, then quenched leaving substitutionally dissolved sulphur (with
$\Delta V < 0$) in solution, and stressed. Fig. 4 shows the sulphur coverage
on fracture surfaces, eqn. (12), as a function of temperature predicted
for the average crack propagation velocity $v = 0.1 \mu ms^{-1}$ and the crack
residence time $\Delta t = 1s$, using an empirical relation between crack
growth rate and stress intensity, and appropriate values of the
material parameters involved [4,5]. We note excellent agreement with
the measured data over the whole range of coverage, corresponding to
0.2 - 20% monolayer sulphur deposition on, for instance, a (100)
crystallographic plane.

Experimental evidence strongly supports the basic fracture
mechanism. Scanning Auger microprobe analysis indicates that sulphur,

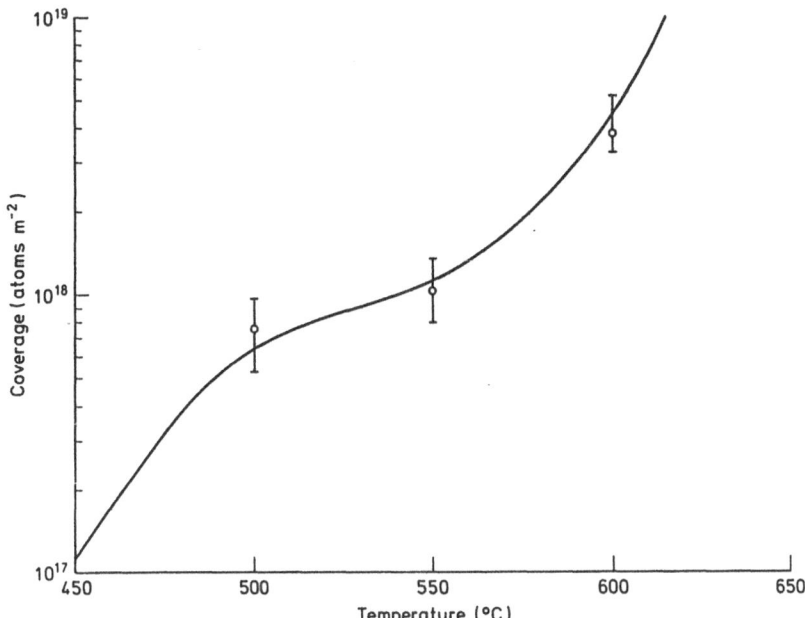

Fig. 4. Theoretically predicted sulphur coverage on fracture surfaces as a function of temperature (——), together with the available measured data (o) and error bars.

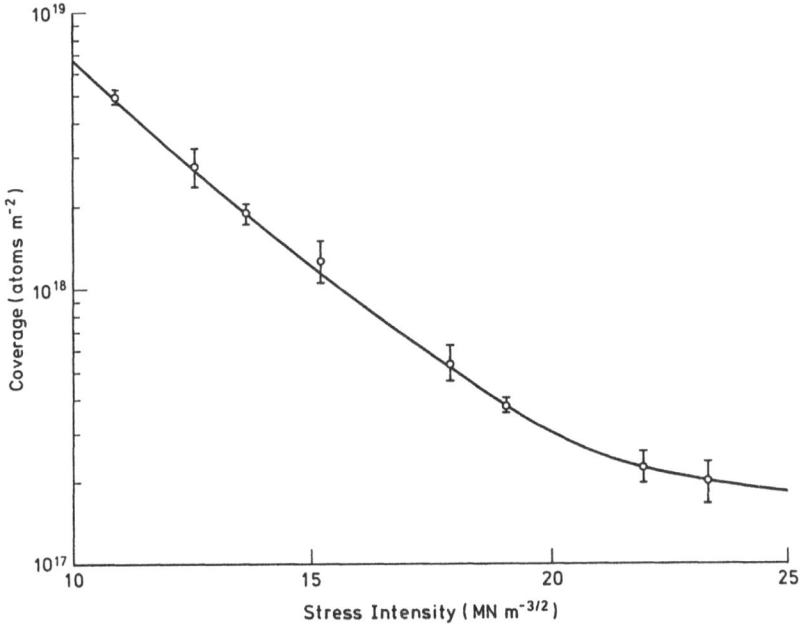

Fig. 5. Theoretically predicted hydrogen coverage on fracture surfaces as a function of stress intensity (——), together with experimental values (o) and error bars deduced from measured data of the external hydrogen gas pressure.

set free during austenitisation, is the principal source of the observed segregated sulphur solute [9]. Scanning electron microscopy and acoustic emission measurements confirm the 'step-wise' nature of the crack growth, the observed average jump distance being consistent with the scale of influence of high tensile stress around crack-tips [10,11]. Finally, optical analysis reveals the meandering nature of the fracture paths, suggesting that mixed-mode loading of the crack-tips prevailed for the most part of their history [5].

b) Hydrogen-induced brittle intergranular fracture
This type of cracking commonly occurs in high-strength steels; it represents one of the most widely documented examples of environment-sensitive response of iron-based materials, caused by hydrogen diffusing to the locally enhanced stress field near crack-tips and acting there synergistically with other embrittling grain-boundary segregants such as those quoted above [12,13].

Consider a notched bend specimen of NiCr steel under stress exposed to a gaseous hydrogen atmosphere at room temperature supplying hydrogen atoms interstitially dissolved (with $\Delta V > 0$) in the iron lattice. Fig. 5 shows the predicted hydrogen coverage on fracture surfaces, eqn. (12), as a function of threshold stress intensity for the onset of crack growth. The coverage is deduced by converting measured data of the external hydrogen gas pressure relating to this stress intensity [14], with $v = 1 \mu ms^{-1}$ and $\Delta t = 1s$, upon deriving the concentration of hydrogen in solution and using appropriate values for the material parameters concerned [4,15]. Unfortunately, there are as yet no means of measuring the hydrogen coverage itself. However, the qualitative variation displayed in Fig. 5 agrees with behaviour expected from observations of the segregation kinetics of more quantifiable solutes (e.g. sulphur or phosphorus) whose efficacy in reducing the local stress intensity rises with the enrichment of grain-boundaries [11,16]. The range of coverage, corresponding to 0.4 - 14% monolayer hydrogen deposition on a (100) plane, implies an embrittling potency comparable to that of the aforementioned solutes, in accord with quantum mechanical investigations [17], while scope for interaction with these elements remains.

Evidence again supports our model hypotheses. The experimental conditions ensure that a constant background hydrogen concentration is maintained from which hydrogen depletes [14]. Scanning electron micro-scopy analyses here too reveal a 'step-wise' crack propagation growth, and intergranular fracture surface morphologies suggest mixed-mode loading of the crack-tips for the most part of their history [18].

Conclusion

We have presented a kinetic model of solute segregation near crack-tips during intergranular fracture of materials, based on stress intensity and temperature. This admits transport of solute atoms to the crack as well as to the grain-boundary ahead of the crack (the latter feature being thought a prerequisite for embrittlement to take place within the fracture process envisaged) when the segregating point-defects are undersized or oversized and mixed-mode loading prevails. The model predictions and inherent hypotheses have been correlated with observations of two particular fracture phenomena in ferritic steels and, unlike other suggestions (see, e.g., [19,20]), receive support from experimental and theoretical evidence to date. Moreover, interpretation of these phenomena in terms of the proposed fracture mechanism is quantitatively consistent with the overall phenomenon of enrichment of fracture zones due to the stress-driven segregation of embrittling solutes.

The theoretical results have a concise analytic form owing to two essential simplifications, viz. the assumption that the actual stress field is that of an ideal sharp crack in a purely elastic body, and the neglect of random diffusion of the point-defects. Finite element calculations allowing for tip blunting and plasticity however indicate total solute flows very much like those in the elastic case, but with the origin of the flow lines centred at the position of maximum hydrostatic stress, displaced by about twice the crack root radius in front of the crack, rather than at the physical crack-tip [5]. Random diffusion, which would tend to smooth out the transient point-defect concentration distributions, has been estimated to have merely little effect on the number of solute segregating in the relevant (short) time of crack arrests between crack jumps [9]. Such diffusion processes have been included recently in the case of an isolated crack, when its tip only acts as a sink for the migrating point-defects [21].

One final point concerns the fact that the addressed segregation and concomitant grain-boundary fracture events are not specific to the chemical nature of a particular solute or host matrix in hand. Thus, whilst the above examples clearly demonstrate the importance of stress-driven segregation of embrittling solutes as a controlling factor for intergranular cracking in a certain class of steels, future research may well find this mechanism operating in other materials as well.

I am grateful to Drs R. Bullough, FRS and C.A. Hippsley for discussions and comments. The present work was supported by the UKAEA Programme on Underlying Research.

References

1. A.R. Troiano, *Trans. Am. Soc. Met.* 52, 54 (1960).
2. E. Lunarska, *Acta Metall.* 26, 1805 (1978).
3. C.A. Hippsley, J.F. Knott and B.C. Edwards, *Acta Metall.* 30, 641 (1982).
4. C.A. Hippsley, H. Rauh and R. Bullough, *Acta Metall.* 32, 1381 (1984).
5. H. Rauh, C.A. Hippsley and R. Bullough, *Acta Metall.* 37, 269 (1989).
6. H. Rauh and R. Bullough, *Proc. R. Soc. London* A397, 121 (1985).
7. H. Rauh and R. Bullough, *Proc. R. Soc. London* A427, 1 (1990).
8. C.J. McMahon Jr, in *Advances in Fracture Research*, p. 143. Pergamon Press, Oxford, England (1984).
9. C.A. Hippsley, *Acta Metall.* 35, 2399 (1987).
10. C.A. Hippsley, D.J. Buttle and C.B. Scruby, *Acta Metall.* 36, 441 (1988).
11. P. Bowen and C.A. Hippsley, *Acta Metall.* 36, 425 (1988).
12. D.P. Williams and H.G. Nelson, *Metall. Trans.* 1, 63 (1970).
13. C.A. Hippsley and N.P. Haworth, *Mater. Sci. Technol.* 4, 791 (1988).
14. R.A. Oriani and P.H. Josephic, *Acta Metall.* 22, 1065 (1974).
15. C.A. Hippsley and C.L. Briant, *Scr. Metall.* 19, 1203 (1985).
16. J. Kameda and C.J. McMahon Jr, *Metall. Trans.* 12A, 31 (1981).
17. R.P. Messmer and C.L. Briant, in *Hydrogen Degradation of Ferrous Alloys*, p. 140. Noyes Publications, Park Ridge, New Jersey (1985).
18. J. Kameda, *Acta Metall.* 34, 1721 (1986).
19. I.W. Chen, *Acta Metall.* 34, 1335 (1986).
20. J.P. Hirth, *Metall. Trans.* 11A, 861 (1980).
21. F.R. Brotzen and A. Seeger, *Acta Metall.* 37, 2985 (1989).

CONTINUUM MECHANICS OF MEDIA WITH INTERFACES

Ladislav V. Berka
Institut of Theoretical and Applied Mechanics
Czechoslovak Academy of Sciences
Vyšehradská 49, 128 49 Prague 2

Abstract. The basis for the description of the mechanical properties of materials is the definition of a model representing the characteristic quality of real materials. The first step in the analysis of the quality of any mechanical system is the definition of a geometrical model, represented in the branch of materials by a structural model. With respect to their structure materials can be divided into two groups, namely simple, with one physical /atomic or molecular/ level of a structure, and complex, with the structure on a level of particles and continuous or discontinuous phases.
This paper is concerned with a model of non-homogeneous materials with interfaces, i.e. with the polycrystalline structure. It derives quantities describing materials with volume and surface inhomogeneities and shows the procedure for deriving the equations of continuum mechanics for materials with such a structure.

Introduction. All natural, technologically processed and man-made composite materials have a complex internal structure which significantly influences their mechanical properties and behaviour under load. With regard to the considerable variety of structural forms and, consequently, the variety of their influence on the properties of materials it is necessary to analyse this problem in a greater detail.

The starting point for a theoretical description of the mechanical properties of a material is the choise of the model displaying its quality. The first step in the analysis of the properties of any mechanical system is the definition of a geometrical model, represented, in the field of materials, by a structural model.

For the purpose of describing their structure, materials can be divided into two groups, namely simple materials, with one physical level of structure /atomic or molecular/, and complex materials, represented by structures consisting of a mixture of homogeneous particles or continuous and discontinuous phases respectively.

The theory of physical structures of solids, crystallography /1/, is based on point models of simple solid substances, while the theory of particle structures lacks such universal structural objects. For exam-

ple, the model of a quasi-homogeneous or heterogeneous material using
the volume ratio of phases, neglects such inhomogeneities as the interfaces, edges and vertices of polycrystalline as well as other grains.
The theory of structures of this type forms the subject of "stereology"
/2/, which defines their stereometric models and invariant dimensions,
i.e. structural parametres.

The deformation of a mechanical system is understood as a relative change of the distance between two adjacent points, which can be of a local
or global character. In relation to deformation characteristics of materials, we shall therefore call the deformation of a structural model
a deformation model of a material. The present work is concerned with
the model of the material with interfaces. It presents a procedure for
deriving structural parameters and equations of continuum mechanics /3/
for media with interfaces.

Structure of polycrystals. Among complex materials, polycrystals represent a considerably large group of materials, a typical structural object of which is a crystaline grain having, in an idealized case, the
shape of an irregular polyhedron. Besides of the volume of the grain,
also its surfaces, edges and vertices represent important objects of the
polycrystalline structure, on whose number per unit volume of the complex material also its resulting mechanical properties depend /4/. Such
a structural model of a polycrystal corresponds also to the fundamental stereological quantities and relations /Table 1/,studied earlier in
metallography /5/,/6/ with purpose of their measurements and determination of their correlations with mechanical properties.

Table 1. Stereological quantities and their interrelations

		Spatial features		
Volume	Surface	Line	Point	Dimension
V_V =	A_A =	L_L =	P_P	mm^{o}
	S_V =	$4L_A/\pi$ =	$2P_L$	mm^{-1}
		L_V =	$2P_A$	mm^{-2}
	$P_V S_V/2 = 2P_A P_L$ =		P_V	mm^{-3}

To ensure the possibility of introduction of these objective parameters
also into the theoretical description of the mechanical properties of
polycrystals, i.e. their constitutive equations, it is necessary to define them analytically. In the author´s paper /7/, it is shown, how it

is possible to represent the differentials of geometric quantities of a polycrystalline structure in an analytical space by means of a local affine transformation. Let

$$X_I = X_I (x_k) \quad , \qquad \det\left[\frac{\partial x_I}{\partial x_k}\right] \neq 0 \quad , \tag{1}$$

be an unequivocal representation defining the structure of materials in the form of the relation between the Lagrange X_K and the Euler x_j coordinates. The differential element of the polycrystalline structure is considered in the form of parallelepiped, for the structural components of which the differentials of edge lengths, surface areas and volume contents, the following expressions hold in Lagrange and Euler coordinates respectively, see /Fig.1/,

$$d\,X_I \quad , \quad d\,S_I = \varepsilon_{IJK} dX_J^{\backprime} dX_K^{\backprime\backprime} \quad , \quad d\,V = 1/3\ dX_I\ dS_I \ ,$$

$$\tag{2}$$

$$d\,x_i \quad , \quad d\,s_i = \varepsilon_{ijk} dx_j^{\backprime} dx_k^{\backprime\backprime} \quad , \quad d\,v = 1/3\ dx_i\ ds_i \ ,$$

where ε_{IJK} and ε_{ijk} are the Levi-Civita antisymmetric unit tensors.

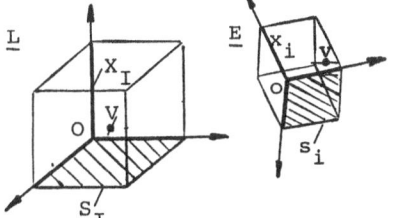

Figure 1. Differentials of structure components in Lagrange and Euler coordinate systems.

Between the differentials of coordinates dX_I and dx_i, representing the edge components of the structure, the following transformation relations hold

$$d\,X_I = X_{I,j}\ d\,x_j \quad , \quad X_{I,j} = \frac{\partial x_I}{\partial x_j} \quad ,$$

$$\tag{3}$$

$$d\,x_i = x_{i,J}\ d\,X_J \quad , \quad x_{i,J} = \frac{\partial x_i}{\partial x_J} \quad ,$$

and further pay the identities

$$X_{I,j}\ x_{j,K} = \delta_{IK} \quad , \quad x_{i,K}\ X_{K,j} = \delta_{ij} \quad , \tag{4}$$

where δ_{ij} and δ_{IJ} are Kronecker's unit tensors.

Between the differentials of surface structure components dS_I and ds_i hold analogously the following transformation relations /7/

$$d\,S_I = S_{I,j}\ d\,s_j \quad , \quad S_{I,j} = \frac{\partial s_I}{\partial s_j} \quad ,$$

$$\tag{5}$$

$$d\,s_i = s_{i,J}\ d\,S_J \quad , \quad s_{i,J} = \frac{\partial s_i}{\partial s_J} \quad ,$$

and further pay again identities

$$s_{I,j}\, s_{j,K} = \delta_{IK} \ , \quad s_{i,J}\, s_{J,k} = \delta_{ik} \ . \qquad (6)$$

The differentials of volume structure components transform themselves according to the relations

$$d\,V = \frac{\partial V}{\partial v}\, dv \ , \quad \frac{\partial V}{\partial v} = X_{I,j}\, S_{I,j} = \det \lfloor X_{I,k} \rfloor . \quad (7)$$

The representation of the structure of a material requires the quantitative preservation of all structural components in the given macrovolume of the material. Therefore, for the representation of volume structure components from the microvolume element of a material into the volume element of an analytical space, shown in figure 2, the following relation hold

$$\sum_{1}^{n} \int_{v^{(n)}} d\,v^{(n)} \ \longrightarrow \ \int_{V} (\sum_{1}^{n} \partial v^{(n)} / \partial V)\, d\,V \qquad (8)$$

If $\partial v^{(n)}/\partial V$ is written in the form of (7) and the summation according to (n) is made alternately over $S_{I,j}$ and $X_{I,j}$, we obtain the mutual relations

$$(X_{I,j}\, S_{I,j})^{(n)} = (X_{I,j})^{(n)}\, S_{I,j} = X_{I,j}\, (S_{I,j})^{(n)}. \qquad (9)$$

If we put $X_I \equiv x_j$, then $X_{I,j} = S_{I,j} = \delta_{Ij}$ and after rearrengement we obtain those relations in the form

$$\partial v^{(n)} / \partial V \ = \ \partial s^{(n)} / \partial S \ = \ \partial x^{(n)} / \partial X \ , \qquad (10)$$

whose meaning is identical with the first line of Table 1 and expresses the equivalence óf the volume, area and linear measurements of the quota of volume structure components in a macro-volume of material.

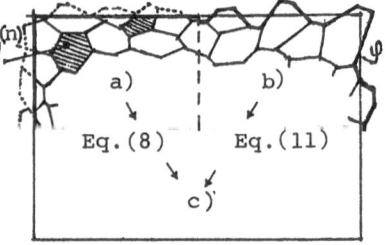

Figure 2. The representation of a) volume and b) surface structure components in the analytical space c)

The preservation of the inner surface structure component in the macro-volume considered, follows from the following hypothesis. The macroscopic elementary parallelepiped with polycrystalline structure is extracted from the material by a cut leading along the structural surfaces nearest to the macro-surface of the parallelepiped so that its resulting content of the volume remains preserved - as it is shown in figure 2. If the integral over the structural surface is changed into the integral over the macro-surface, using the transformation relation between the differentials of surfaces according to the Eq.(5)

$$\int_{\mathscr{I}} d\ s_i \quad \longrightarrow \quad \int_S s_{i,I}\ d\ S_I \quad , \qquad \mathscr{I} = V \quad , \qquad (11)$$

and, further, using Gauss integral theorem, is changed into the volume integral

$$\int_S s_{i,J}\ d\ S_J \ = \ \int_V \frac{\partial^2 s_i}{\partial x_I \partial s_I} d\ V \ = \ \int_V \frac{\partial s_i}{\partial v}\ d\ V \ , \qquad (12)$$

then the term following the last integral expresses the specific inner surface and is identical with the first parameter in the second line of Table 1, i.e. S_V .

Deformation of polycrystals. The differences between monocrystals and polycrystals, as far as their structure and properties, are very well known. Having in mind a chemically pure metal, the structure of poly-crystal is then characterized by a great number of individual grains with probability distribution of orientation of crystallographic axis and of a size. The studies of dependence of mechanical properties of polycrystals on grains orientation, especially elastic moduli, have rich bibliography enough /8/. They were started by Voigt /9/ and Reuss /10/ and in the last twenty years they were advanced by Hashin /11/ and Kröner /12/. The studies of grain size influence on strength and yield stress of polycrystals were started by Zener /13/, Hall /14/ and Petch /15/ in the middle of this century and are not satisfactorily solved out till now, as seen from some of the latest papers /16/,/17/. This situation follows from the fact that the presented problem is formula-ted as geometrical and not as thermodynamical, according to its nature. The polycrystal is a thermodynamic system of particles described by in-ternal parameters of the partial volume and and the interface area.

Mechanical properties of materials are influenced by their structure through the micro-deformation mechanism which manifests itself by the non-homogeneity of microstresses and microdeformation fields. It is in the interest of the study of the laws gowerning the formation of mecha-nical properties of materials to analyse these fields. On the basis of the works done by Rowinskij /18/ and Dawson /19/, in which small kine-matic rotations of grains of a polycrystal under elastic behaviour were proved roentgenographically and microscopically, the author /20/ desig-ned a mechanical model of a polycrystal with such properties. To verify previous works and, on the other hand, to obtain approximate quantita-tive strain and rotation values of polycrystalline grains, author /21/ executed by a micro-photogrammetry method new experiment which proved the last results and gave new ones. A photoelastic modelling was also used for verifying of the mechanism of grain rotation /22/. An element of a continuum with grain boundaries as an inherent property is a new mecha-

nical model in the analysis of this problem. The model of a polycrystal
with idealized grain-boundaries, which has been designated the "ideal
polycrystal" model, consists of elastic grains with normal bonds acting
along them. The results of the stress analysis in bonds show, that there is a rotation of the grain, since the stresses along the grain edges
change and give moments to the centre of the grain. This result shows
that the grain-boundary possessing properties interpreted by the suggested model, and the grain asymmetry, are the basis of the mechanism
of grain rotation. Thus, rotations observed in a polycrystal during its
deformation, may be assumed to be the result of the same structure and
mechanism.

Continuum mechanics of structured media. The basic structural model in
the contemporary mechanics of materials is the point model of a solid
substance derived from the physical reality of the atomic lattice of metals. In an analytical description this model is represented by the concept of a homogeneous continuum. From the range of real materials only
materials with a single, monocrystalline structural level correspond
accurately with this model. The deformation of point structures is understood as their affine transformation, the coefficients of which are represented by the strain tensor components. The deformation model is then
the deformation of a homogeneous parallelepiped produced by a homogeneous stress field /23/. The deformation characteristics of simple materials /elastic/ thus depend only on physical quantities, such as the
absolute temperature. Generalization of the deformation model - by the
incorporation of the effects of flexure and torsion - results in the
linearized case of Cosserat continuum, i.e. the Mindlin - Tiersten theory of polar elasticity, represented by the equations /24/, writen in Lagrange coordinate system. Here express

$$P_I = T_{IJ} n_J \ , \quad Q_I = M_{IJ} n_J \ , \quad \text{the boundary conditions,} \quad (13)$$

$$T_{IJ,J} + F_I = 0 \ , \quad \varepsilon_{IJK} T_{KJ} + M_{IJ,J} + G_I = 0 \ , \quad \text{the eqs.} \quad (14)$$
$$\text{of equilibrium,}$$

$$E_{IJ} = \frac{1}{2} (U_{I,J} + U_{J,I}) \ , \quad \text{the strain tensor,} \quad (15)$$

$$K_{IJ} = \frac{1}{2} \varepsilon_{JKL} U_{L,KI} = R_{J,I} \ , \quad \text{the bending-torsion tensor,} (16)$$

$$K_{JI,L} = K_{LI,J} \ , \quad \varepsilon_{LHJ}(E_{IJ,H} + \varepsilon_{KIH} K_{JK}) = 0 \ , \quad \text{the eqs.} \quad (17)$$
$$\text{of compatibility,}$$

$$\delta L = \frac{1}{2} \int_S (dP_I \delta U_I + dQ_I \ \delta R_I) =$$
$$= \frac{1}{2} \int_V (T_{IJ} \ \delta E_{IJ} + M_{IJ} \ \delta K_{IJ}) \, d V \ , \quad \text{the virtual work.} \quad (18)$$

where P_I, O_I, are vectors of force and moment stresses, T_{IJ}, M_{IJ}, are the tensors of force and moment stresses, F_I, G_I, are unit volume forces and moments, U_I is the displacement vector, R_I is rotation vector, E_{IJ} is the small strain tensor and K_{IJ} is the bending-torsion tensor.

Continuum mechanics of "ideal polycrystal". The theory of the deformation of granular quasihomogeneous structures /8/ has been based so far, as shown in the preceeding paragraphs, on the geometry and properties of volume structure components and their spatial arrangement and has neglected the influence of other components such as the surface structure. The derivation of the equations of mechanical behaviour of a granular structure continuum represents the problem of transformation of the equations of continuum mechanics introduced before. These are written in Lagrange coordinates and derivation of those for the continuum with interfaces is carried out by means of the representation in Euler coordinate system in which the transformation coefficients $x_{i,J}$ are identical with the structural parameters. It is in contradistinction to the non-linear theory of elasticity, in which these transformation coefficients are represented by the displacement gradients.

However, between the differantials of the vectors of stresses and displacements in the Lagrangian and Eulerian coordinate systems, the transformation relation

$$d\,P_I = X_{I,k}\,d\,p_k \,, \qquad d\,U_I = X_{I,j}\,d\,u_j \,, \qquad (19)$$

hold analogously. With the use of Eqs. (3) and (5), following relations arise from Eq. (19), (13), (5),

$$T_{IJ} = X_{I,k}\,s_{j,J}\,t_{kj} \,, \qquad M_{IJ} = X_{I,k}\,s_{j,J}\,m_{kj} \,. \qquad (20)$$

After the insertion of the terms from Eq. (20) into the Eqs. (14), differentiation and arrangement we obtain a system of equations describing the equilibrium state of the continuum with interfaces.

$$(X_{I,j}\,s_{k,J})_{,J}\,t_{jk} + X_{I,j}\,|x_{j,K}|\,t_{jk,k} = 0 \,, \qquad (21)$$

$$\qquad\qquad\qquad\qquad\qquad\qquad\qquad\qquad\qquad\qquad\qquad (22)$$

$$\varepsilon_{IJK}\,X_{K,j}\,s_{k,J}\,t_{jk} + (X_{I,j}\,s_{k,J})_{,J}\,m_{jk} + X_{I,j}\,|x_{j,k}|\,m_{jk,k} = 0.$$

The virtual change of the differential of the displacement $d\,U_I$, Eq. (19), have two parts

$$\delta(d\,U_I) = \delta(X_{I,j})\,d\,u_j + X_{I,j}\,\delta(d\,u_j) \,, \qquad (23)$$

the first of which represents a change of an amount of the structure / interfaces / and the second one, the change of the usual strain gradient. The full change of the macrogradient is then expressed by a formula

$$\delta(U_{I,J}) = (X_{I,ji} \, u_{i,k} \, \delta x_j + u_{i,k} \, X_{I,j} \, \delta x_{j,i} + X_{I,i} \, \delta u_{i,k}) x_{k,J}$$

and   in which mean $X_{I,ji}$, $x_{j,i}$, the change of the struc-   (24)
ture quantities and $u_{i,k}$ is elastic microdeformation. We shall show
further the equations of compatibility, Eqs.(17), transformed into the
Eulerian coordinate system, when elastic microdeformation is assumed
only .

$$\varepsilon_{ijh} (c_{1j} \, k_{i1,h} + c_{11} \, k_{ih,j}) = 0 \qquad (25)$$

$$\varepsilon_{ijm} \, \varepsilon_{nj1} \left[c_{mp} \, e_{pg,i} + \tfrac{1}{2} \, \varepsilon_{mgs} \, \varepsilon_{qhs} \, \varepsilon_{qop} \, c_{hp} \, k_{io} + \right.$$
$$\left. + c^{PI}_{mg} (X_{I,p} \, x_{q,P})_{,i} \, e_{pq} + \tfrac{1}{2} \, \varepsilon_{mgh} \, \varepsilon_{sqn} \, \varepsilon_{pqo} \, S_{J,h} (s_{s,J} \, c_{np})_{,i} \, r_o \right] = 0 \qquad (26)$$

We have denoted here $c_{ij} = X_{I,i} \, X_{I,j}$ the metric tensor of a structure, $c^{KL}_{ij} = 1/2 (X_{K,i} \, X_{L,j} + X_{L,i} \, X_{K,j})$ is a fourth order tensor of the
structure. The e_{pq} is the strain tensor, k_{ij} is the bending-torsion
tensor and r_o is the rotation vector in Euler - structure - coordinates. The quantities c_{ij} and c^{KL}_{ij} enter into the theory as orderd values.

For virtual work according to Eq.(18), after transformation and arrangement with the use of equations of equilibrium, Eqs.(21),(22), we obtain the expression

$$\delta L = \int_V \left\{ t_{ij} \, c_{jk} \, \delta e_{ki} + \tfrac{1}{2} |X_{L,i}| \, m_{ij} \, \varepsilon_{nh1} \, [\varepsilon_{ipl} \, b_{pn} \, \delta k_{jh} + \right.$$
$$\left. + \varepsilon_{1pq} \, s_{i,I} \, (b_{np} \, S_{I,q})_{,j} \, \delta r_h] \right\} d \, v. \qquad (27)$$

If we now compare the form of the equations for continua with granular
/polycrystalline/ structure with the equations for a homogeneous continuum, we find the following differences. First, there is an explicit
dependence of structural /Eulerian/ stresses and strains on structural
charakteristics. Further, there is a difference between the equations
of equilibrium and equations of compatibility in the Lagrangian and the
Eulerian /structural/ coordinate system. The Eqs.(21) and (22),in which
the body forces and body moments are not assumed, contain against the
Eqs. (14) yet the parts with t_{ij} and m_{ij}, multiplied by gradients of
structural parametres. The equations of compatibility contain also similar parts and these afford the oportunity of introducing into the solution, both in the equation of moment stresses equilibrium, and in the
second compatibility equation, the conditions which will enable the description of the kinemetic motion /local rotations/ of parts of the structure. The structural continuum will then be incompatible on its own level, eventhough the macroscopic compatibility will be complied with.
The procedure which is shown here can be concluded also into the des-
description of fracture phenomena, if we shall suppose that macrocrack
is straight, but microscopically quasistraight, in a sense of grainboundary.

LITERATURE REFERENCES

/1/ Kittel, Ch., Introduction to Solis State Physics, Chapter 1,
J. Wiley, New York /1956/

/2/ Underwood, E.E., Quantitative Stereology, Chapter 1, Addison-Wesley
Publ. Comp. Massachusetts /1970/

/3/ Berka, L., Acta Stereologica,Vol. 6, No.1./1987/p. 95-102

/4/ Saxl, I., Stoyan, D., Metallic Materials, 23, 3 /1985/, p.298-307

/5/ Saltykov, A.S., Stereometric metallography, Metallurgizdat,
Moscow /1970/

/6/ Stoyan, D., Mecke, J. Stochastische Geometrie, Akademie-Verlag,
Berlin /1983/

/7/ Berka, L., Acta Technica 5, p. 568-581, Czechoslovak Acad. Sci.
/1972/

/8/ Shermergor, T.D., Theory of Elasticity of Micro-non-homogeneous
Media, Izd. Nauka, Fiziko-Mat. Lit., Moscow/1977/

/9/ Voigt, W., Lehrbuch der Kristallphysik, Teubner, Leipzig und
Berlin /1910/

/10/ Reuss, A., Zeitschrift für Angew. Math. und Mech., 9, 49,/1929/

/11/ Hashin Appl. Mech. Rev. 17, No.1, p.1 /1964/

/12/ Kröner, Z., Phys. 151, No. 4, p. 504 /1958/

/13/ Zener, C., Phys. Rev. 69, p. 128 /1945/

/14/ Hall, E.O., Proc. Phys. Soc. London, B 64, p. 747 /1951/

/15/ Petch, N.J., J. Iron Steel Inst., 174, p. 25 /1953/

/16/ Beevers, C.J., Fatigue Thresholds, Emas Wasley, 1., p.257 /1982/

/17/ Lukáš, P., Kunz, L., Metallic Materials /Kovové materiály/ 25,
5, p. 622-637 /1987/

/18/ Rovinskij, B.M., Sinajskij, V.M., Some Problems of Strength of
Solids, USSR Acad. Press Moscow /1958/

/19/ Dawson, T.H., PhD. Thesis, The John Hopkins University,
Baltimore /1968/

/20/ Berka, L., J. Mat. Sci. 17 /1982/, p. 1508 - 1512

/21/ Berka, L., Růžek, M., J. Mat. Sci. 19 /1984/, p. 1486-1495

/22/ Berka, L., Acta Technica 2, p.147-160, Czechoslov.Acad.Sci./1985/

/23/ Green, A.E., Zerna, W., Theoretical Elasticity, Ch.2., Clarendon
Press, Oxford /1954/

/24/ Nowacki, W., Theory of Elasticity, Ch.11, Warsaw /1970/

FRACTURE OF COATED PLATES AND SHELLS UNDER THERMAL SHOCK

F. Erdogan and A. A. Rizk
Lehigh University, Bethlehem PA 18015

Abstract. The main interest in this study is in the subcritical crack propagation and fracture of coated materials under repeated thermal shock. First it is shown that the crack problem for a cylindrical shell may be approximated by a plate on an elastic foundation. Then the thermal shock problem for a layered plate supported by an elastic foundation and containing two cracks of arbitrary sizes and locations is considered. An additional factor studied is the effect of surface cooling rate on the stress intensity factors at the crack tips.

Introduction

For the purpose of analyzing the mechanics of the problem, in applications many engineering components may locally be represented by a "plate" or a "shell." In some cases these components consist of a base material or a substrate and a coating. Thermal barrier coating of super alloys by ceramics used in jet engines, stainless steel cladding of nuclear pressure vessels, and a great variety of diffusion bonded materials used in microelectronics may be mentioned as some examples. Typically, these materials are subjected to severe residual stresses upon cooling from their processing temperatures. During the operation they may also undergo certain thermal cycling. Depending on the temperature gradients, the underlying thermal stress problem may be treated either as a thermal shock problem or as a quasistatic isothermal problem in the sense that the problem may still be time-dependent but with no variation of temperature within the composite solid.

Usually an important mode of mechanical failure in such materials is the subcritical crack growth due to creep or low cycle fatigue. The cracks generally start from micro-flaws near or at the interface, or at the surface and grow perpendicularly to the nominal interface. For the service life assessment of the component, aside from the subcritical crack growth and fracture characterization of the material, what is needed is the determination of the stress intensity factor as a function of the crack length and the time.

In this paper a number of unique aspects of the problem will be discussed. First, to make the composite shell problem analytically tractable it will be shown that the crack problem for the shell can be modelled by a plate on an elastic foundation. The two dimensional composite medium containing cracks perpendicular to the interface will then be solved under thermal shock conditions. The second question to be studied will be the examination of the influence of the cooling rate on the stress intensity factors in thermal shock problems. The results will be given for a surface crack, a crack terminating at or intersecting the interface, and a crack initiating from the interface.

The Temperature Distribution

Subject to verification, for the purpose of analyzing the thermal shock problem, in this paper the relatively thin composite cylindrical shell will be approximated by a layered plate on an elastic foundation. The procedure is to obtain the transient temperature distribution by solving the diffusion equations for the composite plate problem, to use this temperature distribution in a layered plate on an elastic foundation in the absence of cracks to determine the stress distribution, and finally by applying the equal and opposite of this transient thermal stresses to the crack surfaces to solve the crack problem. Referring to Fig. 1, consider the temperature distribution in the composite medium which is initially at a homogeneous temperature T_∞ and at $t \geq 0$ and $x = 0$ is subjected to the thermal boundary condition shown.

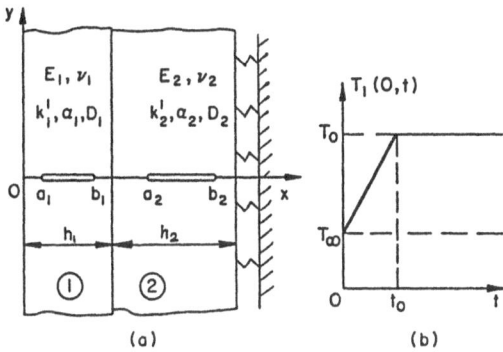

Fig. 1   The crack geometry and the temperature boundary condition.

Defining

$$T_i(x,t) - T_\infty = \theta_i(x,t) , \quad (i=1,2) ,$$ (1)

the problem may be formulated as follows:

$$\frac{\partial^2 \theta_i}{\partial x^2} = \frac{1}{D_i} \frac{\partial \theta_i}{\partial t} , \quad (i=1,2) ,$$ (2)

$$\theta_i(x,0) = 0 , \quad (i=1,2) ,$$ (3)

$$\theta_1(h_1,t) = \theta_2(h_1,t) ; \quad k_1' \frac{\partial \theta_1(h_1,t)}{\partial x} = k_2' \frac{\partial \theta_2(h_1,t)}{\partial x} ,$$ (4)

$$\frac{\partial}{\partial x} \theta_2(h_1 + h_2,t) = 0 ,$$ (5)

$$\theta_1(0,t) = \theta_0 H(t) \ , \quad \theta_0 = T_0 - T_\infty \ , \tag{6a}$$

$$\theta_1(0,t) = \frac{\theta_0}{t_0}\left[tH(t) - (t-t_0)H(t-t_0)\right] \ , \quad \theta_0 = T_0 - T_\infty \ , \tag{6b}$$

where k_i' and D_i , (i=1,2), are the coefficients of heat conduction and the thermal diffusivity, respectively. The solution of the problem may be obtained in a rather straightforward manner by using Laplace transforms and the residue theorem [1], [2]. Thus, by defining the dimensionless quantities

$$x' = x/h_1 \ , \quad m = (D_1/D_2)^{\frac{1}{2}} \ , \quad \eta = mk_1'/k_2' \ , \quad \tau = tD_1/h_1^2 \ ,$$

$$\tau_0 = t_0 D_1/h_1^2 \ , \quad \gamma = mh_2/h_1 \ , \tag{7}$$

for the boundary condition (6b) we find

$$\frac{\theta_1(x,t)}{\theta_0} = \frac{\tau}{\tau_0} + 2\sum_{n=1}^{\infty}(e^{-\tau\lambda_n^2} - 1)*$$

$$* \frac{\cos\lambda_n(x'-1)\cos\lambda_n\gamma + \eta\sin\lambda_n(x'-1)\sin\lambda_n\gamma}{\tau_0\lambda_n^3[(1+\eta\gamma)\sin\lambda_n\cos\lambda_n\gamma + (\eta+\lambda)\cos\lambda_n\sin\lambda_n\gamma]} \ , \quad 0 < x' < 1 \ , \quad \tau \leq \tau_0 \ ,$$

$$\frac{\theta_1(x,t)}{\theta_0} = 1 - 2\sum_{n=1}^{\infty}e^{-\tau\lambda_n^2}(e^{\tau_0\lambda_n^2}-1)*$$

$$* \frac{\cos\lambda_n(x'-1)\cos\lambda_n\gamma + \eta\sin\lambda_n(x'-1)\sin\lambda_n\gamma}{\tau_0\lambda_n^3[(1+\eta\gamma)\sin\lambda_n\cos\lambda_n\gamma + (\gamma+\eta)\cos\lambda_n\sin\lambda_n\gamma]} \ , \quad 0 < x' < 1 \ , \quad \tau > \tau_0 \ ,$$

$$\tag{8}$$

$$\frac{\theta_2(x,t)}{\theta_0} = \frac{\tau}{\tau_0} + 2\sum_{n=1}^{\infty}(e^{-\tau\lambda_n^2}-1)*$$

$$* \frac{\cos[\lambda_n(x'-1-h_2/h_1)m]}{\tau_0\lambda_n^3[(1+\eta\gamma)\sin\lambda_n\cos\lambda_n\gamma + (\eta+\gamma)\cos\lambda_n\sin\lambda_n\gamma]} \ , \quad 1 < x' < 1 + \frac{h_2}{h_1} \ , \quad \tau < \tau_0 \ ,$$

$$\tag{9}$$

$$\frac{\theta_2(x,t)}{\theta_0} = 1 - 2\sum_{n=1}^{\infty}e^{-\lambda_n^2\tau}(e^{\lambda_n^2\tau_0}-1) *$$

$$* \frac{\cos[\lambda_n(x'-1-h_2/h_1)m]}{\tau_0\lambda_n^3[(1+\eta\gamma)\sin\lambda_n\cos\lambda_n\gamma + (\gamma+\eta)\cos\lambda_n\sin\lambda_n\gamma]} \ , \quad 1 < x' < 1 + \frac{h_2}{h_1} \ , \quad \tau > \tau_0 \ ,$$

where λ_n , $n = 1,2,\ldots$ are the roots of

$$\cos\lambda_n\cos\gamma\lambda_n - \eta\sin\lambda_n\sin\gamma\lambda_n = 0 \ . \tag{10}$$

The solution for the boundary condition (6a) is obtained by letting $\tau_o \rightarrow 0$ in (9) and (10).

Thermal Stresses in the Uncracked Plate

Referring again to Fig. 1, if the uncracked infinite composite plate is elastically supported, it will remain flat under the selfequilibrating transient thermal stresses. Using the symmetry conditions and the Hooke's law, and integrating through the thickness, the stresses due to the temperature distribution (8) and (9) may be expressed as (see [2] for details)

$$\sigma_{1yy}^T(x,t) = \frac{E_1}{1-\nu_1} [\epsilon_o(t) - \alpha_1' \theta_1(x,t)] , \quad 0 < x < h_1 , \tag{11}$$

$$\sigma_{2yy}^T(x,t) = \frac{E_2}{1-\nu_2} [\epsilon_o(t) - \alpha_2' \theta_2(x,t)] , \quad h_1 < x < h_2 , \tag{12}$$

$$\epsilon_o(t) = \frac{E_1 \alpha_1' h_1 (1-\nu_2) \bar{\theta}_1(t) + E_2 \alpha_2' h_2 (1-\nu_1) \bar{\theta}_2(t)}{E_1 h_1 (1-\nu_2) + E_2 h_2 (1-\nu_1)} , \tag{13}$$

where

$$\bar{\theta}_1(t) = \frac{1}{h_1} \int_0^{h_1} \theta_1(x,t) dx , \quad \bar{\theta}_2(t) = \frac{1}{h_2} \int_{h_1}^{h_1+h_2} \theta_2(x,t) dx . \tag{14}$$

In the axisymmetric deformations of a cylindrical shell due to the resistance provided by curvature to transverse displacements, the shell having a mean radius R_n and thickness h may be approximated by a plate on an elastic foundation having a stiffness $\chi = Eh/R_n^2$ [2], [3]. Thus, for a composite shell it may easily be shown that

$$\chi = (E_1 h_1 + E_2 h_2)/R_n^2 \tag{15}$$

where R_n is the radius of the neutral circle and may be determined from simple equilibrium considerations as follows:

$$R_n = R_i + \frac{h_1}{2} + \frac{E_2 h_2 (h_1 + h_2)}{2(E_1 h_1 + E_2 h_2)} . \tag{16}$$

The Crack Problem

To determine the self equilibrating transient thermal stresses the cylindrical shell is approximated by a fully constrained flat plate, the transverse constraints coming from an elastic support. Since the plate is infinite, in the uncracked plate the in-plane stresses σ_{yy} and σ_{zz} are independent of the stiffness of the support. In the crack problem, however, there will be local bending and the stresses will depend on the stiffness of the support. Thus, for the problem described in Fig. 1, Navier's equations are solved, in addition to the standard continuity, regularity, and boundary conditions, under

$$\sigma_{2xy} = 0 , \quad \sigma_{2xx} + \chi u_2 = 0 , \quad x = h_1 + h_1 , \quad y = 0 \tag{17}$$

$$\sigma_{1yy}(x,0) = -\sigma_{1yy}^T , \quad a_1 < x < b_1 ; \quad v_1(x,0) = 0 , \quad 0 < x < a_1 , \quad b_1 < x < h_1 , \tag{18}$$

$$\sigma_{2yy}(x,0) = -\sigma_{2yy}^T , \quad a_2 < x < b_2 ; \quad v_2(x,0) = 0 , \quad h_1 < x < a_2 , \quad b_2 < x < h_1 < h_2 . \tag{19}$$

By defining the unknown functions

$$\phi_i(x) = \frac{\partial}{\partial x} v_i(x,+0) , \quad (i=1,2) , \tag{20}$$

the problem may be reduced to the following system of singular integral equations (see [2] for details)

$$\int_{a_1}^{b_1} [\frac{1}{s-x} + k_{11}(x,s)]\phi_1(s)ds + \int_{a_2}^{b_2} k_{12}(x,s)\phi_2(s)ds = - \frac{\pi(1+\kappa_1)}{4\mu_1} \sigma_{1yy}^T(x,t) , \quad a_1 < x < b_1 ,$$

$$\tag{21}$$

$$\int_{a_1}^{b_1} k_{21}(x,s)\phi_1(s)ds + \int_{a_2}^{b_2} [\frac{1}{s-x} + k_{22}(x,s)]\phi_2(s)ds = - \frac{\pi(1+\kappa_2)}{4\mu_2} \sigma_{2yy}^T(x,t) , \quad a_2 < x < b_2 ,$$

$$\tag{22}$$

subject to the single-valuedness conditions (if $a_1 > 0$, $b_1 \neq a_2$, $b_2 < h_1 + h_2$)

$$\int_{a_i}^{b_i} \phi_i(s)ds = 0 , \quad (i=1,2) . \tag{23}$$

Referring to [4], the general solution of (21) and (22) may be expressed as

$$\phi_j(s) = \frac{g_j(s)}{(s-a_j)^{\alpha_j}(j-s)^{\beta_j}} , \quad a_j < s < b_j , \quad 0 < Re(\alpha_j,\beta_j) < 1 , \quad (j=1,2) , \tag{24}$$

where g_j is unknown and is H-continuous in $a_j \leq s \leq b_j$, $(j=1,2)$. The asymptotic analysis shows that the kernels k_{ij} , $(i,j=1,2)$ can be expressed as

$$k_{ij}(x,s) = k_{ij}^f(x,s) + k_{ij}^s(x,s) , \quad (i,j=1,2) , \tag{25}$$

where k_{ij}^f , are bounded in the corresponding closed intervals $a_i \leq x \leq b_i$, $a_j \leq s \leq b_j$, $(i,j=1,2)$ and k_{ij}^s may have end point singularities. For example, it can be shown that

$$k_{11}^s(x,s) = - \frac{1}{s+x} + \frac{6x}{(s+x)^2} - \frac{4x^2}{(s+x)^3} + \frac{c_{11}}{2h_1-x-s} + \frac{c_{12}(h_1-x)}{(2h_1-x-s)^2} + \frac{c_{13}(h_1-x)^2}{(2h_1-x-s)^2} ,$$

$$a_1 < (x,s) < b_1 , \tag{26}$$

where c_{11} , c_{12} , and c_{13} are known bi-material constants [5]. Similar expressions for the other kernels and details of the asymptotic analysis are given in [2].

Thus, in the limiting cases of $a_1 = 0$, $b_1 = h_1$, $a_2 = h_1$, and $b_2 = h_1 + h_2$ the end point singularities of the kernels would have an influence on the singular behavior of the functions ϕ_1 and ϕ_2 and the characteristic equations giving α_i and β_i, (i=1,2) may be obtained by using the function theoretic method [5], [4]. After determining α_i and β_i the functions g_i, (i=1,2) could be found by following [6] and [7]. The stress intensity factors at the crack tips may then be defined by and evaluated from

$$k_1(a_j) = \lim_{x \to a_j - 0} \sqrt{2(a_j - x)}\, \sigma_{jyy}(x,0) = \frac{4\mu_j}{1+\kappa_j} \frac{g_j(a_j)}{\sqrt{(b_j - a_j)/2}} \, , \quad a_1 > 0 \, , \, a_2 > h_1 \, , \, (j=1,2)$$

$$(27)$$

$$k_1(b_j) = \lim_{x \to b_j + 0} \sqrt{2(x - b_j)}\, \sigma_{jyy}(x,0) = -\frac{4\mu_j}{1+\kappa_j} \frac{g_j(b_j)}{\sqrt{(b_j - a_j)/2}} \, , \quad b_1 < h_1 \, , \, b_2 < h_1 + h_2 \, ,$$

$$(j=1,2) \qquad (28)$$

$$k_1(h_1) = \lim_{x \to h_1 + 0} \sqrt{2}(x - h_1)^{\beta_1}\, \sigma_{2yy}(x,0) = \frac{4\mu_2}{1+\kappa_2} \frac{\sqrt{2}\, g_1(h_1)(d_{21} + \beta_1 d_{22})}{(h_1 - a_1)^{\alpha_1} \sin\pi\beta_1} \, , \quad b_1 = h_1 \, , \, a_2 > h_1 \, ,$$

$$(29)$$

$$k_1(h_1) = \lim_{x \to h_1 - 0} \sqrt{2}(h_1 - x)^{\alpha_2}\, \sigma_{1yy}(x,0) = \frac{4\mu_1}{1+\kappa_1} \frac{\sqrt{2}\, g_2(h_1)(d_{11} - d_{12}\alpha_2)}{(b_2 - h_1)^{\beta_2} \sin\pi\alpha_2} \, ,$$

$$b_1 < h_1 \, , \, a_2 = h_1 \, , \qquad (30)$$

where the bimaterial constants d_{ij}, (i,j=1,2) are defined in [5]. Similar expressions may be developed for the stress intensity factors $k_{xx}(h_1)$ and $k_{xy}(h_1)$ at the singular point $x = h_1$, $y = 0$ [5].

Results and Discussion

First to show the significance of the elastic support in dealing with crack problems, the results of the plane strain edge crack problem for a homogeneous plate under uniform tension σ_0 in y direction with or without the elastic support are given in Table 1. In this case we have $h_1 + h_2 = h$, $E_1 = E_2 = E$, $\nu_1 = \nu_2 = \nu$, $a_1 = 0$, and $b_1 = b$ (see Fig. 1). Note that the constraint provided by the support can reduce the stress intensity factor quite considerably.

Regarding the main contention of this paper that a cylindrical shell with an axi-symmetric circumferential crack may be approximated by a plate on an elastic foundation, some results comparing the stress intensity factors resulting from a uniform tension are given in Table 2. Similar results for the transient thermal stresses are shown in Table 3. It may be seen that the approximation is, in fact, fairly good.

Some typical calculated results showing the stress intensity factors in a layered plate on an elastic foundation as a function of the dimensionless time τ (the Fourier number) are given in Figures 2-9. The thermoelastic properties of materials

TABLE 1. Normalized stress intensity factors $k_1(b)/\sigma_o\sqrt{b}$ in a plate containing an edge crack and subjected to uniform tension σ_o under plane strain conditions without and with elastic support. $\chi h/E = 0.01108.$ χ is the stiffness of the support, h is the plate thickness and b is the crack depth.

b/h	$\chi = 0$	$\chi \neq 0$
0.0	1.1215	1.1215
0.1	1.1892	1.1574
0.2	1.3673	1.2725
0.3	1.6599	1.4267
0.4	2.1114	1.5983
0.5	2.8246	1.8135
0.6	4.0332	2.0775
0.7	6.3549	2.3801
0.8	11.955	2.7399
0.85	18.628	2.9835
0.9	34.632	3.3660

TABLE 2. Comparison of the normalized stress intensity factor in a uniformly stressed plate with an edge crack under plane strain conditions supported by an elastic foundation of stiffness χ with that in a cylindrical shell containing an internal circumferential crack and subjected to a uniform axial stress. The shell results are obtained for $R_i/h = 9$ in [8]. The stiffness of the foundation is $\chi = Eh/R_n^2$, $R_n = R_i + h/2$, or for $R_i/h = 9$, $\chi h/E = 0.01108$.

b/h	$k_1(b)/\sigma_o\sqrt{b}$ Shell	$k_1(b)/\sigma_o\sqrt{b}$ Plate	% Error
0.1	1.158	1.157	-0.09
0.2	1.253	1.272	1.49
0.3	1.392	1.426	2.44
0.4	1.568	1.598	1.91
0.5	1.779	1.813	1.91
0.6	2.025	2.077	2.57

TABLE 3. Comparison of the normalized stress intensity factor $k_1(b)/\sigma_o^T\sqrt{b}$ in a plate on an elastic foundation containing an edge crack under plane strain conditions with that in a cylindrical shell containing an internal circumferential crack. In both cases the external load is the transient thermal stresses resulting from a step change θ_o in the wall temperature. $R_i/h = 9$, $\chi h/E = 0.01108$, $\tau = tD/h^2$, $\sigma_o^T = -E\alpha\theta_o/(1-\nu)$.

b/h	$\tau = 0.01$ Shell	$\tau = 0.01$ Plate	$\tau = 0.05$ Shell	$\tau = 0.05$ Plate	$\tau = 0.1$ Shell	$\tau = 0.1$ Plate	$\tau = 0.5$ Shell	$\tau = 0.5$ Plate
0.01	0.962	0.957	0.833	0.821	0.724	0.709	0.277	0.261
0.1	0.657	0.653	0.701	0.691	0.633	0.621	0.247	0.233
0.2	0.426	0.432	0.589	0.592	0.560	0.559	0.224	0.218
0.3	0.300	0.307	0.501	0.509	0.502	0.507	0.206	0.200
0.4	0.238	0.240	0.432	0.435	0.453	0.437	0.190	0.183
0.5	0.205	0.207	0.378	0.379	0.408	0.408	0.174	0.168
0.6	0.185	0.187	0.334	0.337	0.367	0.367	0.158	0.153

used in the examples are given by the following table:

Material Pair	k_2'/k_1'	D_2/D_1	α_2'/α_1'	E_2/E_1	ν_2/ν_1	R_i/L	$\chi L/E_2$
A	3	3	0.75	1	1	9	0.01108
B	3.385	4.070	2.294	0.611	1	9	0.01185

Material pair A corresponds to a stainless steel layer (mat. 1) welded on a ferritic steel base simulating the cladded pressure vessels. The Pair B represents a ceramic layer (mat. 1) bonded to a steel substrate (mat. 2). The normalizing stress intensity factor used in the figures is given by $\sigma_o^T\sqrt{\ell}$, where $\sigma_o^T = -E_1\alpha_1'\theta_o/(1-\nu_1)$ and ℓ is the crack length. The results given are largely self-explanatory.

Figure 3 clearly shows the effect of the loading rate as measured by τ_o on the stress intensity factors. In Figures 5-9 the crack either terminates at or crosses the interface. In Fig. 5, since $E_1 > E_2$, at the crack tip we have a stress singularity β_1 that is greater than 1/2, whereas for the same reason in Figures 7 and 8 α_1 is less than 1/2. Figure 9 shows a sample result for the stress intensity factor k_x in a crack crossing the interface defined by $k_{x\cdot} = \lim_{y\to+0} y^\beta \sigma_{xx}(h_1,y)$.

References

1. H. S. Carslaw and J. C. Jaeger, Conduction of Heat in Solids, Oxford University Press, 1950.

2. A. A. Rizk, "Cracking of a Layered Medium on an Elastic Foundation under Thermal Shock," Ph.D. Dissertation, Lehigh University, 1988.

3. S. Timoshenko and J. N. Goodier, Theory of Elasticity, McGraw-Hill, New York, 1951.

4. N. I. Muskhelishvili, Singular Integral Equations, P. Noordhoff, N.V., 1953.

5. A. A. Rizk and F. Erdogan, J. of Thermal Stresses, 12, 169-189, 1989.

6. A. C. Kaya and F. Erdogan, Q. Appl. Math. 45, 105-122, 1987.

7. A. C. Kaya and F. Erdogan, Q. Appl. Math. 45, 455-469, 1987.

8. H. F. Nied and F. Erdogan, Inst. J. of Fracture, 22, 277-301, 1983.

9. H. F. Nied and F. Erdogan, J. of Thermal Stresses, 6, 1-14, 1983.

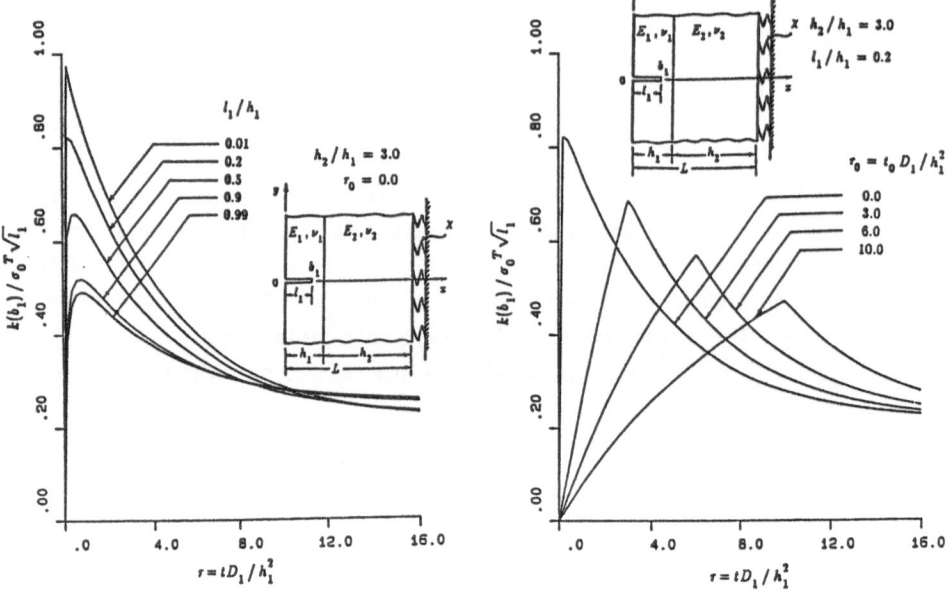

Fig. 2. $k_1(b_1)$ vs. τ (material pair A)   Fig. 3. $k_1(b_1)$ vs. τ (material pair A)

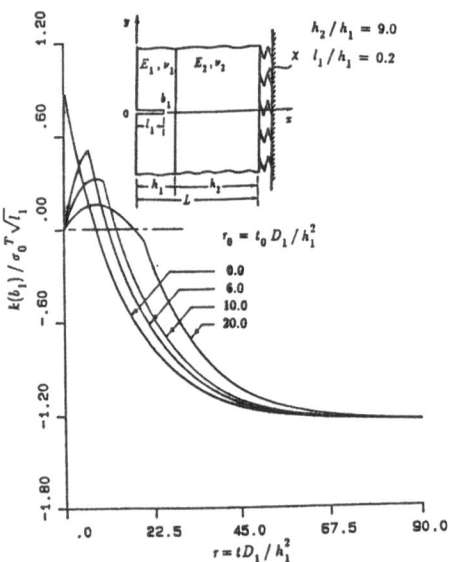

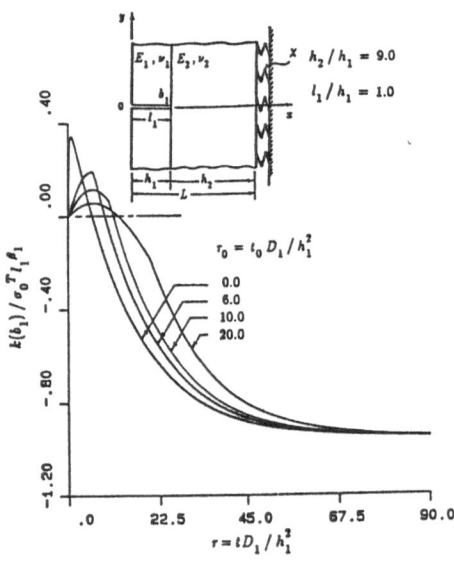

Fig. 4. $k_1(b_1)$ vs. τ (mat. pair B)

Fig. 5. $k_1(b_1)$ vs. τ (mat. pair B),
$b_1 = h_1$, $\beta_1 = 0.552538$

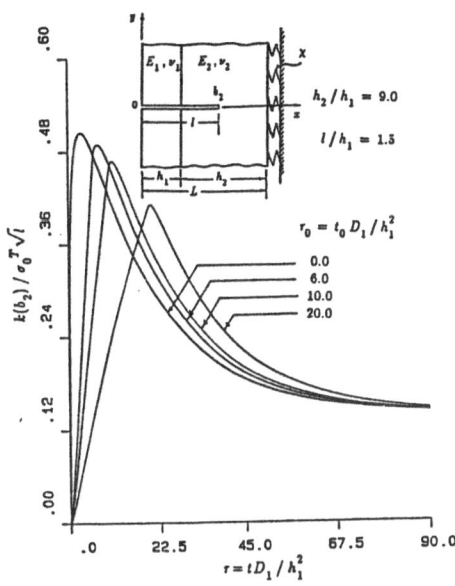

Fig. 6. $k_1(b_2)$ vs. τ , (mat. pair A)

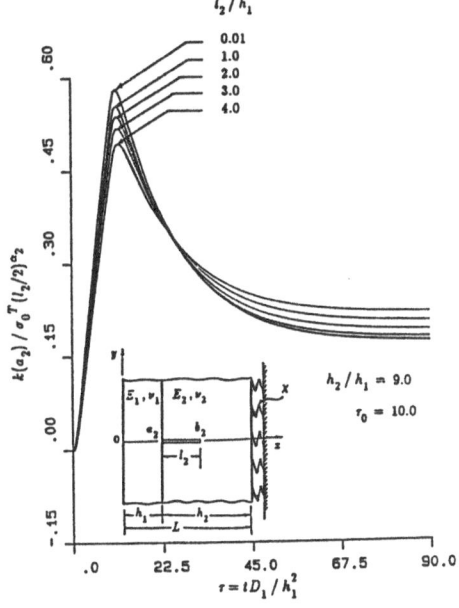

Fig. 7. $k_1(a_2)$ vs. τ (mat. pair B),
$a_2 = h_1$, $\alpha_2 = 0.451242$

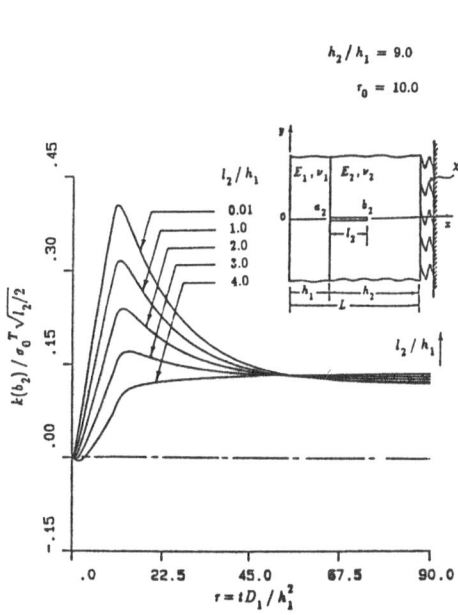

Fig. 8. $k_1(b_2)$ vs. τ (mat. pair B),
$a_2 = h_1$, $\alpha_2 = 0.451242$

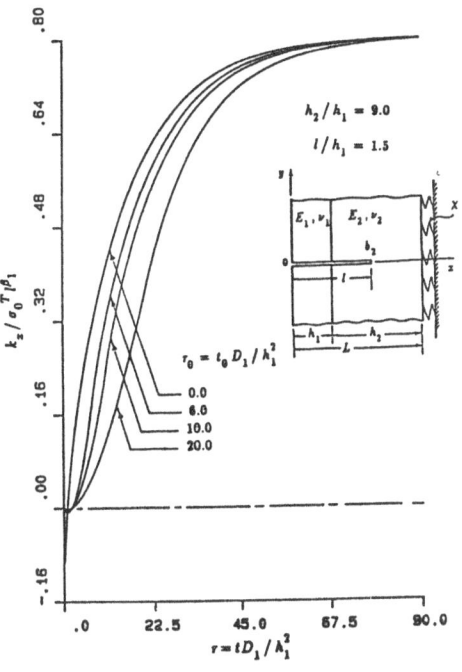

Fig. 9. $k_x(h_1)$ vs. τ (mat. pair B),
$b_1 = a_2$, $\beta_1 = \alpha_2 = \beta = 0.0187224$

Debonding of thin surface layers generated by thermal and diffusive fluxes

Zbigniew S. Olesiak

Faculty of Mathematics, Informatics and Mechanics,
University of Warsaw, Palace of Culture and Science,
00-901 Warsaw, Poland

1. Fundamental equations

The hardening of surface layers is frequently necessary for technological reasons. Coated solids get the required properties and become resistant to friction. The coatings, however change the material properties, in the result the solid is no longer homogeneous. In the case of sharp changes in the values of material constants the redistribution of stresses, arising from a heat flux or (and) diffusive mass flux can be significant. The effect of thin coating of a material with high Young's modulus is particularly pronouncing. In the limit cases one can assume that the layer is inextensible. This assumption let us make the use of the methods and notions of the fracture mechanics.

The system of partial differential equations of the diffuso-thermo-elasticity will serve as a point of departure. In the stationary case we obtain the following system of equations

$$(1-2\nu)\nabla^2 \mathbf{u} + \text{grad div } \mathbf{u} = 2(1+\nu)[\alpha_\theta \text{ grad}\theta + \alpha_c \text{grad}c], \qquad (1.1)$$

$$\nabla^2 \theta = 0, \quad \nabla^2 c = 0, \qquad (1.2)$$

where $\mathbf{u}$ denotes the displacement vector, θ is the deviation of temperature with respect to that of the natural state T, c denotes the diffusing mass concentration, μ, λ are Lamé's constants, ν - Poisson's ratio, α_θ, α_c - material constants,

The constitutive equations, the generalized Duhamel-Neumann relationships, take the form

$$\sigma = 2\mu\varepsilon + (\lambda \text{ tr}\varepsilon - \gamma_T \theta - \gamma_c c)\mathbf{1}, \qquad (1.3)$$

while the remaining constitutive equations read

$$s = \gamma_T \text{tr }\varepsilon - d_1 c + m\theta, \quad M = -\gamma_c \text{tr }\varepsilon + d_1 \theta + ac. \qquad (1.4)$$

Here σ and ε are the stress and strain tensors, respectively, $\mathbf{1}$ is the unit tensor, s - the entropy, M - the chemical potential, γ_T, γ_c, d_1, a, m are material constants.

An important problem in the mechanics of solids is to find the distribution

of stresses, and particularly the stress concentration. In the diffuso-termo-
elasticity we wish to know whether such a stress concentration, or per-
haps the regions of stress singularities can be generated by the fields of
temperature and/or that of mass diffusion. The purpose of the paper is to
determine the distribution of stresses in solids to which thin, inextensible
coatings are bonded. It can be assumed that either the diffusing mass can
penetrate through the membrane or that the membrane itself is its source.

2. The boundary value problems for solids with inextensible coatings

We assume that the bending rigidity of an inextensible membrane is
negligibly small. Consequently we can assume that there is no normal
stress component exerted by thermal or diffusive effects. If a solid is of
layered type, i.e. a semispace or a layer, and its bounding plane is coated
in the entire region then the shear stress components depend on the character of the distribution of heat, and diffusion fluxes over the bounding
membrane. They will be continuous functions provided heat and diffusion
fluxes are sufficiently smooth. On the other hand if the bounding plane is
coated only on its part singularities exist in the distribution of the shear
stress components, and the character of the singularities does not depend
on the distribution of heat and/or diffusion fluxes.
Consequently we obtain that the tangential components of the displacement
vector vanish over the entire bonding. The normal component of the displacement vector is to be determined from the solution of the boundary value
problem. The second mechanical boundary condition states that the normal
stress component is zero. In general case the analytical solution is not
available. In the case of the axial symmetry the problem has all the features of the general three-dimensional one, and the solution can be determined analytically in terms of special functions. Therefore we shall discuss
the axially symmetric cases.
The system of partial differential equations (1.1)-(1.2) in the cylindrical
system of coordinates $\varkappa = (r, \vartheta, z)$ with axially symmetric displacement vektor $\boldsymbol{u} = (u, 0, w)$ assumes the following form

$$2(1-\nu) B_1 u + (1-2\nu) Du + \partial/\partial r D(ru) = 2(1+\nu)\partial/\partial r(\alpha_\Theta \Theta + \alpha_c c),$$

$$(1-2\nu) B_0 w + 2(1-\nu) D^2 w + r^{-1}\partial/\partial r D(ru) = 2(1+\nu)D(\alpha_\Theta \Theta + \alpha_c c), \qquad (2.1)$$

$$(B_0 + D^2)\Theta(r, z) = 0, \quad (B_0 + D^2)c(r, z) = 0,$$

where we have employed the symbols denoting the differential operators

$$B_k \equiv \frac{\partial}{\partial r^2} + r^{-1}\frac{\partial}{\partial r} - k r^{-2}, \quad k = 1, 2; \ D \equiv \frac{\partial}{\partial z}.$$

In the case when the range of $r \in (0, \infty)$ the system of the partial differential equations (2.1) can be reduced, by means of the Hankel integral trans-

forms of zero and first order, to the one of the ordinary differential equations

$$[(1-2\nu)D^2 - 2(1-\nu)\xi^2]\hat{u} - \xi D\hat{w} = -2(1+\nu)\xi(\alpha_\theta\hat{\Theta} + \alpha_c\hat{c}),$$

$$[2(1-\nu)D^2 - (1-2\nu)\xi^2]\hat{w} + \xi D\hat{u} = 2(1+\nu)D(\alpha_\theta\hat{\Theta} + \alpha_c\hat{c}),$$ (2.2)

$$(D^2 - \xi^2)\hat{\Theta} = 0, \quad (D^2 - \xi^2)\hat{c} = 0,$$

where

$$\hat{u}(\xi,z) = H_1[u(r,z);r\to\xi] = \int_0^{\infty} ru(r,z)J_1(r\xi)dr,$$

$$\hat{w}(\xi,z), \hat{\Theta}(\xi,z), \hat{c}(\xi,z) = H_0[w(r,z),\Theta(r,z),c(r,z);r\to\xi] =$$

$$= \int_0^{\infty} r\{w(r,z),\Theta(r,z),c(r,z)\}J_0(r\xi)dr.$$

The solution of the system of differential equations (2.2), for $z \geq 0$ is easy to find. If we take into account the Duhamel-Neumann constitutive equations

$$\sigma_{rz}(r,z) = \mu(Du + \partial/\partial r\, w),$$ (2.3)

$$\sigma_{zz}(r,z) = (\lambda+2\mu)Dw + \lambda r^{-1}\partial/\partial r(ru) - (3\lambda+2\mu)(\alpha_\theta\Theta + \alpha_c c),$$

and the boundary conditions

$$\sigma_{zz}(r,0) = 0, \quad u(r,0) = 0,$$ (2.4)

we obtain the following solution of the system of equations (2.1) in terms of the inverse Hankel transforms of the zero and first order:

$$2(1-\nu)u(r,z) = (1+\nu)zH_1[\{\alpha_\theta A(\xi) + \alpha_c B(\xi)\}\exp(-\xi z);\xi\to r],$$

$$2(1-\nu)w(r,z) = -(1+\nu)H_0[\xi^{-1}(1-\xi z)\{\alpha_\theta A(\xi) + \alpha_c B(\xi)\}\exp(-\xi z);\xi\to r],$$

$$(1-\nu)\sigma_{rz}(r,z) = (1+\nu)\mu H_1[(1-\xi z)\{\alpha_\theta A(\xi) + \alpha_c B(\xi)\}\exp(-\xi z);\xi\to r],$$

$$(1-\nu)\sigma_{zz}(r,z) = (1+\nu)\mu z H_0[\xi\{\alpha_\theta A(\xi) + \alpha_c B(\xi)\}\exp(-\xi z);\xi\to r],$$ (2.5)

$$(1-\nu)\sigma_{rr}(r,z) = -(1+\nu)\mu H_0[(4-\xi z)\{\alpha_\theta A(\xi) + \alpha_c B(\xi)\}\exp(-\xi z);\xi\to r],$$

$$\Theta(r,z) = H_0[A(\xi)\exp(-\xi z);\xi\to r],$$

$$c(r,z) = H_0[B(\xi)\exp(-\xi z);\xi\to r].$$

It is evident, from the solution (5), that all the mechanical quantities, depend on the value of the expression $\{\alpha_\theta A(\xi) + \alpha_c B(\xi)\}$. Parameters $A(\xi)$ and $B(\xi)$ are to be determined from the solutions of the heat conduction equation, and diffusion equation, respectively.

3. The relationships between the contact shear stress and the distribution of temperature and mass diffusion on the bounding plane

In order to find the contact stresses it is not necessary to follow the whole procedure of finding the transforms parameters $A(\xi)$ and $B(\xi)$ with the consecutive inversion of the corresponding Hankel transforms. We are able to do it directly. Let us make use of the formula which can be derived from Eqs. 3, 6, and 7 of the system (2.5). We find the following equations (compare [6,8]):

$$\frac{\partial}{\partial r}[r\sigma_{rz}(r,z)] = \frac{1+\nu}{1-\nu}r\mu H_0[(1-\xi z)\xi\{\alpha_\Theta A(\xi) + \alpha_c B(\xi)\}\exp(-\xi z);\xi \to r], \quad (3.1)$$

and

$$D\{\alpha_\Theta A(\xi) + \alpha_c B(\xi)\} = -H_0[\xi\{\alpha_\Theta A(\xi) + \alpha_c B(\xi)\}\exp(-\xi z);\xi \to r]. \quad (3.2)$$

Comparing the above results in the limit, for $z \to 0$, we find the following relationship:

$$\sigma_{rz}(r,0) = -\frac{1+\nu}{1-\nu}\mu r^{-1}\int_0^r s\frac{\partial}{\partial z}[\alpha_\Theta\Theta(s,z) + \alpha_c c(s,z)]\Big|_{z=0}\,ds. \quad (3.3)$$

This means that we can find the shear, contact, stress component by the integration as shown in Eq. (3.3). The next relationship we find by comparing

$$\frac{\partial}{\partial r}[\alpha_\Theta\Theta(s,0) + \alpha_c c(s,0)] = -H_1[\xi\{\alpha_\Theta A(\xi) + \alpha_c B(\xi)\}\exp(-\xi z);\xi \to r], \quad (3.4)$$

and

$$D[\sigma_{rz}(r,z)]|_{z=0} = -2\frac{1+\nu}{1-\nu}\mu H_1[\xi\{\alpha_\Theta A(\xi) + \alpha_c B(\xi)\};\xi \to r], \quad (3.5)$$

whence we obtain:

$$D[\sigma_{rz}(r,z)]|_{z=0} = 2\frac{1+\nu}{1-\nu}\mu\frac{\partial}{\partial r}[\alpha_\Theta\Theta(r,0) + \alpha_c c(r,0)]. \quad (3.6)$$

Formula (3.6) furnishes the value of the shear stress gradient, the distribution of the stress itself cannot be determined in this way since in order to integrate with respect to z one would have to know the solution. However it is noteworthy that, in the case of Dirichlet's boundary conditions, the right hand side of the equation is known from the boundary conditions, and it is not not necessary to know the solution of differential equations (1.2) in order to find the shear stress gradient.

The formulae of a similar type as above can be obtained for $\sigma_{zz}(r,z)$, and for the sum $\sigma_{rr}(r,0) + \sigma_{\vartheta\vartheta}(r,0)$ [8].

4. Distribution of contact shear stress component in solids with coatings

In the case of continuous coatings the problem reduces to finding the function of two variables $\omega(r,z) = [\alpha_\Theta\Theta(r,z) + \alpha_c c(r,z)]$ from the solution of the equations (1.2) together with the given boundary conditions, and consecutive differentiation and integration as shown in Eq. (3.3). The value of $\omega(r,0)$ denotes either the value of the boundary conditions of Dirichlet's type, or it can be computed from the solution of the differential equations in the other cases (Neumann's, radiation type, or mixed boundary conditions) It is important to know how the contact shear stress behave depending on the smoothness of $\omega(r,0)$ on the boundary. We shall consider three particular cases. In the first case function $\omega(r,0)$ is continuous while its first derivative is discontinuous at $r = a$. In the second case there is a jump in the value of function at $r = a$. In the third case we consider the axially symmetric Dirac's distribution. From the form of equation (3.3) we see that for the distributions of $\omega(r,0)$ such that both the function itself and and its first derivative are continuous the shear stress is also a smooth function. As an example of the first case we take

$$\omega(r,0) = \alpha_\Theta \Theta(r,0) + \alpha_c c(r,0) = \gamma a^{-2}(a^2-r^2)H(a-r), \qquad (4.1)$$

where $H(\cdot)$ is the Heaviside unit distribution. From the solution of Laplace's equation

$$\nabla^2\{\alpha_\Theta\Theta(r,z) + \alpha_c c(r,z)\} = \nabla^2\omega(r,z) = 0 \qquad (4.2)$$

which results from Eqs. (1.2), we obtain [5,9,11]

$$\frac{\partial}{\partial z}\omega(r,z)\big|_{z=0} = -\frac{2\gamma}{a^2}\int_0^\infty J_2(a\xi)J_0(r\xi)\,d\xi =$$

$$= -\frac{4\gamma}{\pi a^2 r^2}\Big[(a+r)E(k) - a^2(a+r)K(k)\Big], \quad k = \frac{2\sqrt{ar}}{a+r}, \qquad (4.3)$$

where $K(\cdot)$ and $E(\cdot)$ denote the complete elliptic integrals of the first and second order. Now, if we substitute the result from (4.3) into Eq. (3.3) we shall obtain the contact shear strain distribution also in the form of the complete elliptic integrals:

$$\sigma_{rz}(r,0) = 2\gamma\mu\frac{1+\nu}{1-\nu}\int_0^\infty \xi^{-1}J_2(a\xi)J_1(r\xi)\,d\xi =$$

$$\frac{2\gamma\mu}{3\pi a^2 r(1-\nu)}\Big[(a-r)(a^2+2r^2)K(k) - (a+r)(a^2-2r^2)E(k), \quad k = \frac{2\sqrt{ar}}{a+r}. \qquad (4.4)$$

It is evident from formula (4.4), and the properties of the complete elliptic integrals that function $\sigma_{rz}(r,0)$ is continuous in the entire region $r\in\langle 0,\infty)$, so is its first derivative. However, at $r = a$ there is a point of inflexion. In the second case we take

$$\omega(r,0) = \alpha_\Theta\Theta(r,0) + \alpha_c c(r,0) = CH(a-r), \qquad (4.5)$$

i.e. with the jump at $r = a$ in the value of $[\omega(r,0)]\big|_{r=a} = C$. Then we obtain

$$\frac{\partial}{\partial z}\{\alpha_\Theta\Theta(r,z) + \alpha_c c(r,z)\}\big|_{z=0} = -Ca\int_0^\infty \xi J_1(\xi a)J_0(\xi r)\,d\xi. \qquad (4.6)$$

The integral in (4.6) can be represented in the form of hypergeometric series which also in this case reduces to complete elliptic integrals (see [3]):

$$\frac{\partial}{\partial z}\omega(r,z)\big|_{z=0} = -\frac{2}{\pi}C\begin{cases}a(a^2-r^2)^{-1}E(r/a), & r<a, \\ r(r^2-a^2)^{-1}E(r/a) - r^{-1}K(r/a), & a<r, \end{cases}$$

$$= -\frac{C}{\pi}\Big[(a-r)^{-1}E(k) + (a+r)^{-1}K(k)\Big]. \qquad (4.7)$$

In this case the contact shear stress assumes the form

$$\sigma_{rz}(r,0) = Ca\mu\frac{(1+\nu)}{(1-\nu)}\int_0^\infty J_1(a\xi)J_1(r\xi)\,d\xi =$$

$$= \frac{2}{\pi}C\mu\frac{1+\nu}{1-\nu}a^{-1}r^{-1}\,a[K(r/a) - E(r/a)], \quad r<a$$

$$= -\frac{C}{\pi}\Big[(a-r)^{-1}E(k) + (a+r)^{-1}K(k)\Big]$$

$$= \frac{1}{\pi} \, C\mu \frac{(1+\nu)}{(1-\nu)} \frac{r+a}{r} \left[\frac{r^2+a^2}{(r+a)^2} K(k) - E(k) \right], \quad k = \frac{2\sqrt{ar}}{a+r}. \qquad (4.8)$$

The distribution of shear stress, given by Eq. (4.8), is shown in Fig.1. The solution for the third case can be obtained in a similar way, or by differentiation of Eq. (4.8) with respect to r and division by $2\pi r$. We obtain

$$\sigma_{rz}(r,0) = C\mu \frac{1}{4\pi^2} \frac{1+\nu}{1-\nu} \frac{1}{a} \left[(a-r)^{-1} E(k) + (a+r)^{-1} K(k) \right]. \qquad (4.9)$$

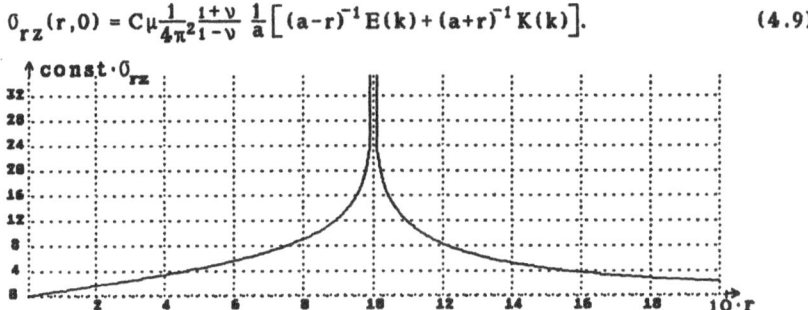

Fig.1 Contact shear stress in case of step function

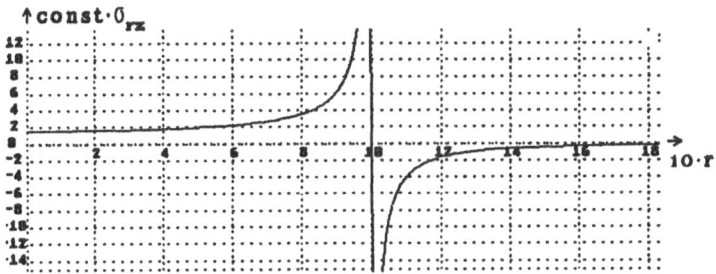

Fig.2 Shear stress for delta distribution

5. A method of solution for discontinuous coating

By discontinuous coating we shall understand finite membranes bonded to the bounding plane. It will be shown that there appear shear stress contration far away from a source of heat and/or mass diffusion. Now, for circular regions of bonding we have, instead of (2.4), the following mechanical boundary conditions on the bounding plane $z = 0$

$$\sigma_{zz}(r,0) = 0, \quad r \in \langle 0, \infty),$$
$$u(r,0) = 0, \quad r \in \langle 0,b), \quad \sigma_{rz}(r,0) = 0, \quad r \in (b,\infty). \qquad (5.1)$$

We assume that the boundary conditions for temperature and mass diffusion are of the form

$$\alpha_\Theta \Theta(r,0) = \Theta_1(r) H(a_1 - r), \quad \alpha_c c(r,0) = \Theta_2(r) H(a_2 - r), \quad a_1 < b, \, a_2 < b. \qquad (5.2)$$

In a special case when $a_1 = a_2 = a$, we can take the sum of the boundary conditions

$$\alpha_\Theta \Theta(r,0) + \alpha_c c(r,0) = \vartheta(r)H(a-r),$$

$$(5.3)$$

corresponding to the equation $\nabla^2[\alpha_\Theta \Theta(r,z) + \alpha_c c(r,z)] = 0$. Since we can make use of the superposition principle it is sufficient to consider the boundary conditions in the form of (5.3).

For mechanical boundary conditions (5.1) the solution of Eqs. (1.1) can be written down in the form

$$u(r,z) = H_1[\xi^{-1}\{-2(1-\nu)+\xi z\}\psi(\xi)+(1+\nu)\big[\alpha_\Theta A(\xi) + \alpha_c B(\xi)\big]\}\exp(-\xi z);\xi \to r],$$

$$\sigma_{rz}(r,z) = 2\mu H_1[(1-\xi z)\psi(\xi)\exp(-\xi z);\xi \to r].$$

$$(5.4)$$

Here $\psi(\xi)$ is the Hankel integral transform parameter which can be determined from the set of dual integral equations, obtained by substitution of $u(r,0)$ and $\sigma_{rz}(r,0)$ into the boundary conditions $(5.1)_2$, and $(5.1)_3$. The form of the solution (5.4) is such that condition $(5.1)_1$ is already satisfied. In this way we obtain:

$$H_1[\xi^{-1}\psi(\xi);\xi \to r] = \frac{1}{2}\frac{1+\nu}{1-\nu}\int_0^\infty \big[\alpha_\Theta A(\xi) + \alpha_c B(\xi)\big]J_1(\xi r)\,d\xi = f(r),\ r\in\langle 0,b),$$

$$(5.5)$$

$$H_1[\psi(\xi);\xi \to r] = 0,\ r\in(b,\infty).$$

The solution of the set of dual integral equations (5.5) is known to be:

$$\psi(\xi) = \sqrt{\tfrac{2}{\pi}}\int_0^b g(s)\sin(s\xi)\,ds,\ \ sg(s) = \sqrt{\tfrac{2}{\pi}}\int_0^s f(t)t^2(s^2-t^2)^{-1/2}\,dt,$$

$$(5.6)$$

and

$$\sigma_{rz}(r,0) = -2\mu\sqrt{\tfrac{2}{\pi}}\frac{\partial}{\partial r}\Big\{\int_r^b g(s)(s^2-r^2)^{-1/2}\,ds\Big\}.$$

$$(5.7)$$

if the thermal and diffusive boundary conditions are in the form (5.2) then

$$rf(r) = \int_0^r \big[\alpha_\Theta \Theta(x,0) + \alpha_c c(x,0)\big]x\,dx.$$

$$(5.8)$$

If we assume that $\vartheta(r)$ in (5.3) is constant, say $\vartheta(r) = C$ then we obtain:

$$g(s) = (2\pi)^{-1/2}\frac{1+\nu}{1-\nu}\mu C\,[s - \sqrt{(s^2-a^2)}H(s-a)],$$

$$\sigma_{rz}(r,0) = -\frac{2}{\pi}\frac{1+\nu}{1-\nu}\mu C\Big[\frac{r}{\sqrt{(b^2-r^2)}}(1-\frac{\sqrt{(b^2-a^2)}}{b}) \; a\frac{2\partial}{\partial r}\int_{\max(r,a)}^b \frac{\sqrt{(s^2-r^2)}}{s^2\sqrt{(s^2-a^2)}}\,ds\Big].$$

$$(5.9)$$

Thus we have obtained the singularity in the value of shear stresses for at the radius of the bonding circle. The singularity is one over the square root of the distance, and its behaviour does not depend on the distribution of the step wise function. The singularities at the jumps of the distribution of temperature or that of the mass diffusion is logarythmic. In Fig. 3 we have shown the diagrams for a = 0.1, 0.5, 0.9, and 1.0. The value C is the same for all curves.

There is another way of getting the solution of the boundary value problem (5.1). Namely we make use of the results discussed at the preceding points, Eqs. (3.3) and the particular case (4.8). The solution then obtained is valid

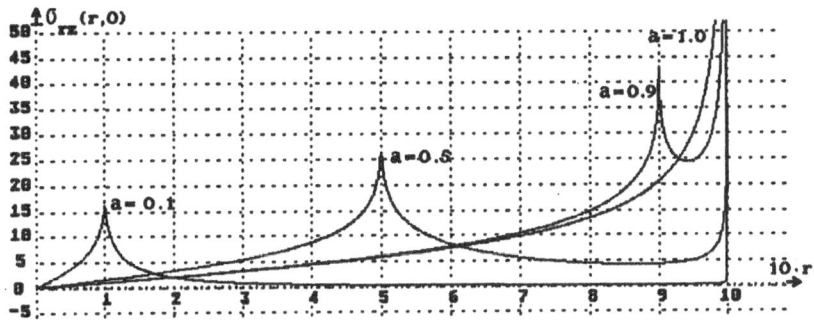

Fig. 3 Contact shear stress distribution for finite bonding $(r = 1.0)$

for a continuous bonding extending over the whole boundary $z = 0$. To obtain the solution for the bonding of finite radius we have to find the distribution of shear stress in the part for $r > b$, and then subtract the solution for an isothermal and isodiffusive case with the same distribution of contact shear stress for $r > b$, however with the opposite sign, and the vanishing radial displacement $u(r,0) = 0$ for $r \in \langle 0,b \rangle$. Thus we obtain the following set of dual integral equations:

$$H_1[\xi^{-1}\psi(\xi); \xi \to r] = 0, \quad r \in \langle 0,b \rangle,$$
$$H_1[\psi(\xi); \xi \to r] = -\sigma_{rz}(r,0), \quad r \in (b,\infty). \tag{5.10}$$

Here $\sigma_{rz}(r,0)$ is the value of shear stress found from Eq.(4.8).

6. The stress intensity factor

The solution of the set of integral equations (5.5) or that of the set (5.10) affords in the result the value of contact shear stress. We see that for a distribution of $\omega(r,0) = \alpha_\theta \Theta(r,0) + \alpha_c c(r,0) > 0$ there appears a singularity of the type corresponding to mode II i.e. in torsion of a solid with a disc shaped crack. Consequently, we can use the methods of fracture mechanics also for problems of bonding of thin inextensible layers. For the particular case discussed at point 5 we have obtained the distribution of contact shear stress. The stress intensity factor is the coefficient which stays at the term with one over square root of the distance from the periphery of the membrane. We have:

$$K_{II} = \frac{2}{\pi} \frac{1+\nu}{1-\nu} \mu \, Cb\left[1 - \frac{\sqrt{(b^2 - a^2)}}{b}\right]. \tag{6.1}$$

If the value K_{II} does not exceed the experimental value $K_{II}{}^0$ then the membrane is bonded to the solid. The initiation of the debonding or cracking occurs when these two values are equal.

Bibliography

1 Adda Y., Philibert J. La Diffusion dans les Solides, Inst. Nat. Sci. et Techn. Nucléaires, Vol. I, II, Paris 1966,

2 Cranck J. The mathematics of diffusion, Clarendon Press, Oxford, second edition, 1975,

3 Eason G., Noble B., Sneddon I.N. On certain integrals of Lipschitz-Hankel type involving products of Bessel functions, Phil. Trans. Roy. Soc. London, Ser. A., Vol. 247, 1955, 529-551,

4 W. Nowacki, Thermoelasticity, 2nd Edition, P.W.N.-Pergamon Press, 1986,

5 H. Olesiak, Z.S. Olesiak, J. Śliżewicz, Stress concentration in elastic coated solids generated by thermodiffusive fluxes, submitted for publication in Surface Physics,

6 Olesiak Z.S. Influence of surface heating on coated elastic solids, J. Thermal Stresses, 12(1989), 293-303,

7 Olesiak Z.S. Stresses in solid bodies exerted by thermodiffusive effects, Teubner Texte zur Mathematik, vol. 111, 217-223,

8 Olesiak Z.S. Stresses due to thermal diffusion in elastic solids with coatings, 3rd Conference on Surface Physics, Inst. Techn. Phys. Mil. Techn. Acad., Warsaw, 1989,

9 Olesiak Z.S. Delaminacja cienkich pokryć i pękanie ośrodka sprężystego na skutek działania efektów termodyfuzyjnych, Zesz. Nauk. Pol. Świętokrzyskiej, Mechanika 45, 1989, 7-23,

10 Oberhettinger F. Tables of Bessel transforms, Springer Verlag, 1972,

11 Прудников А.П., Брычков Ю.А., Маричев О.И. Интегралы и ряды, специальные функции, Наука, Москва, 1983,

12 Sneddon I.N. Special functions of mathematical physics and chemistry, 3rd Ed., Longman Group Lim., 1980.

Acknowledgment

The research was supported by Ministry of National Education of Poland under CPBP 01.08-D2.4.

CRACK PROPAGATION IN MATERIALS WITH LOCAL INHOMOGENEITIES UNDER THERMAL LOAD

A. Bettin, D. Gross

Institut für Mechanik, TH Darmstadt
Hochschulstr. 1, D-6100 Darmstadt

Abstract. The quasistatic crack propagation problem in a plane region with inhomogeneities under thermal loads is investigated. The thermoelastic problem is formulated in terms of singular integral equations which are treated numerically by an appropriate boundary element method. Several examples are discussed. They illustrate the applicability of the solution procedure.

1 Introduction

A crack starts to propagate if certain parameters like stress intensity factors become critical. Then, only in special cases the crack moves along a straight path. In general, depending on the stress field, the crack path will be curved. Especially this will happen, if the body has inhomogeneities like inclusions, holes or interfaces between different materials. Because of the complexity of the problem there are only a few investigations about path predictions in the literature, e.g. [7]. Moreover they all are restricted to pure mechanical loading of the structure.

Similarly, for thermoelastic crack problems only a few analytical solutions are known, all concerning infinite, plane regions with straight cracks [2,3,6]. The problem of a thermally loaded slightly curved crack was analyzed in [1].

In this paper a method of path prediction is proposed for thermally loaded curved cracks, propagating quasistatically in a plane bounded or unbounded region. It bases on an integral equation formulation which leads to a numerical solution procedure in terms of a boundary element method.

2 Integral equation formulation

Using three complex potentials $\Phi(z)$, $\Psi(z)$, $\Theta(z)$ with $z = x + iy$ the field quantities of plane thermoelasticity can be described by the generalized Kolosov equations [2,5]:

$$
\begin{aligned}
\sigma_x + \sigma_y &= 2[\Phi'(z) + \overline{\Phi'(z)}] \\
\sigma_x - \sigma_y + 2i\tau_{xy} &= -2[z\overline{\Phi''(z)} + \overline{\Psi'(z)}] \\
2\mu[u + iv] &= \kappa_1 \Phi(z) - z\overline{\Phi'(z)} - \overline{\Psi(z)} + \kappa_2 \Theta(z) \\
2\alpha_T \mu T &= Re[\Theta'(z)] \\
2\alpha_T \mu[q_x + iq_y] &= -\lambda_T \overline{\Theta''(z)}
\end{aligned}
\tag{1}
$$

$$\text{plane stress} \quad : \quad \kappa_1 = \frac{3-\nu}{1+\nu} \quad , \quad \kappa_2 = 1$$

$$\text{plane strain} \quad : \quad \kappa_1 = 3 - 4\nu \quad , \quad \kappa_2 = 1 + \nu.$$

Herein T denotes the temperature, $q_x + iq_y$ the heat flux vektor, ν the Poisson's ratio, μ the shear modulus, α_T the coefficient of linear thermal expansion and λ_T the thermal conductivity.

We now consider an infinite plane with a traction free curved crack R and an inhomogeneity e.g. a hole with the boundary S. The plane is loaded by a stationary temperature field (Fig. 1).

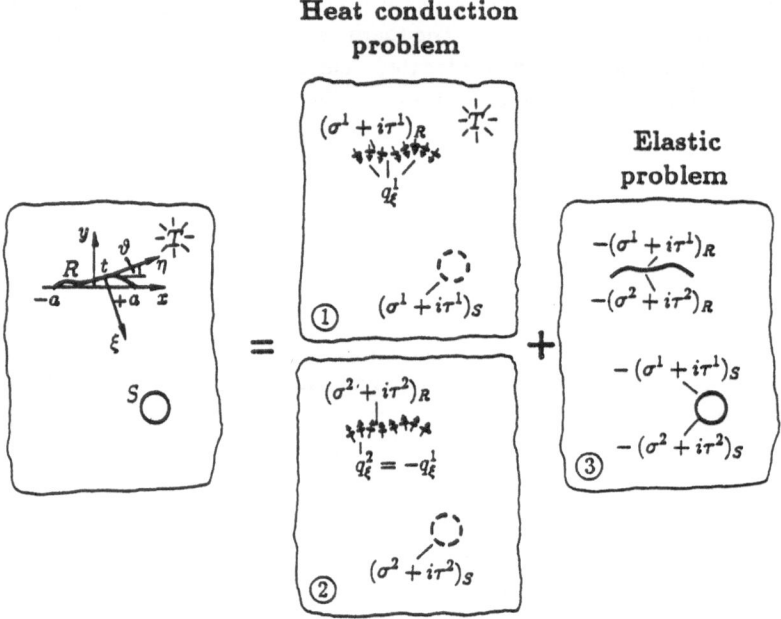

Figure 1: Thermoelastic crack problem

The crack is assumed to act as an isolator while the heat flux shall be undisturbed by the hole. With these assumptions the problem can be solved by the superposition of heat conduction problems and an elastic problem.

In the heat conduction problem 1 the infinite plane without the crack and the hole is loaded by the given stationary temperature field. The accompanied thermostress vector along R, S and the heat flux normal to R are $(\sigma^1 + i\tau^1)$ and q_ξ^1 respectively. In order to fulfill the isolator condition, a heat flux $q_\xi^2 = -q_\xi^1$ along R is prescribed in the problem 2. The resulting stress vector along R, S is $(\sigma^2 + i\tau^2)$.

Because the crack faces R and the boundary S must be traction free, the elastic problem 3 is superposed finally. Here the crack faces and S are loaded by the opposite stresses of the heat conduction problem.

In the following we assume that the heat conduction problem 1 can be solved, so that the stresses $(\sigma^1 + i\tau^1)$ along R and S as well as the heat flux q_ξ^1 are known. If R

is sufficiently smooth, the heat conduction problem 2 can be described by a Fredholm integral equation of first kind for the temperature jump distribution $a(t)$ [1]

$$- q_\xi^1(t) = q_\xi^2(t) = \frac{\lambda_T}{2\alpha_T\mu}\text{Re}\int\limits_R \frac{a(\tau)e^{-i[\vartheta(\tau)-\vartheta(t)]}d\tau}{\tau-t} \qquad \text{with} \qquad \int\limits_R a(s)ds = 0 \quad . \quad (2)$$

From its solution $a(t)$ the dipole distribution

$$\delta(t) = \int\limits_0^t a(\tau)e^{-i\vartheta(\tau)}d\tau \quad , \tag{3}$$

the complex potentials, e.g.

$$\Phi'(z) \;=\; \frac{\kappa_2}{1+\kappa_1}i\int\limits_R \frac{\delta(\tau)d\tau}{\tau-z} \quad , \tag{4}$$

$$\Psi'(z) \;=\; -\frac{\kappa_2}{1+\kappa_1}i\left[\int\limits_R \frac{\delta(\tau)d\bar\tau}{\tau-z} + \int\limits_R \frac{\bar\tau\delta(\tau)d\tau}{(\tau-z)^2}\right]$$

and the field quantities are fully determined [1]. Especially the stress vector along R, S is given by

$$\sigma_\xi^2 + i\tau_{\xi\eta}^2 \;=\; \frac{\kappa_2}{1+\kappa_1}i\left\{2\int\limits_R \frac{\delta(\tau)d\tau}{\tau-t} - \int\limits_R H(t,\tau)\left[\frac{\delta(\tau)d\tau}{\tau-t} + \frac{\delta(\tau)d\bar\tau}{\bar\tau-\bar t}\right]\right\} \tag{5}$$

with

$$H(t,\tau) = 1 - e^{-2i\vartheta(t)}\frac{\tau-t}{\bar\tau-\bar t} \quad .$$

In the pure elastic problem 3 the displacement discontinuities and displacement derivatives at the boundaries R and S respectively can be represented by a dislocation distribution $g(t)$. It can be found as the solution of the singular integral equation [4]

$$p(t) = 2\int\limits_R \frac{g(\tau)d\tau}{\tau-t} - \int\limits_R H(t,\tau)\left[\frac{g(\tau)d\tau}{\tau-t} - \frac{\overline{g(\tau)d\tau}}{\bar\tau-\bar t}\right]$$

$$+ 2\pi i g(t)s + \int\limits_S H(t,\tau)\left[\frac{g(\tau)d\tau}{\tau-t} + \frac{\overline{g(\tau)d\tau}}{\bar\tau-\bar t}\right] \quad , \tag{6}$$

where

$$p(t) = p^1(t) + p^2(t) = \begin{cases} -(\sigma_\xi^1 + i\tau_{\xi\eta}^1)_R - (\sigma_\xi^2 + i\tau_{\xi\eta}^2)_R & \text{on } R \\ -(\sigma_\xi^1 + i\tau_{\xi\eta}^1)_S - (\sigma_\xi^2 + i\tau_{\xi\eta}^2)_S & \text{on } S \end{cases} \tag{7}$$

is known from the heat conduction problem. Because the crack is closed at its ends, the side condition

$$\int\limits_R g(t)dt = 0 \tag{8}$$

must be fullfilled.

From $g(t)$ the complex potentials and all field quantities for this problem can be derived. Especially the stress intensity factors of the whole thermoelastic problem follow from the singular parts of $g(t)$ at the crack tips as shown in the next section.

3   Boundary element method

The numerical solution of the integral equations (2) and (6) is done by a boundary element method using linear isoparametric elements (Fig. 2) with a parameter description for each element as

$$t_l(s) = m_l + \frac{\lambda_l}{2} e^{i\vartheta_l} s \quad , \quad -1 \le s \le +1 \quad . \tag{9}$$

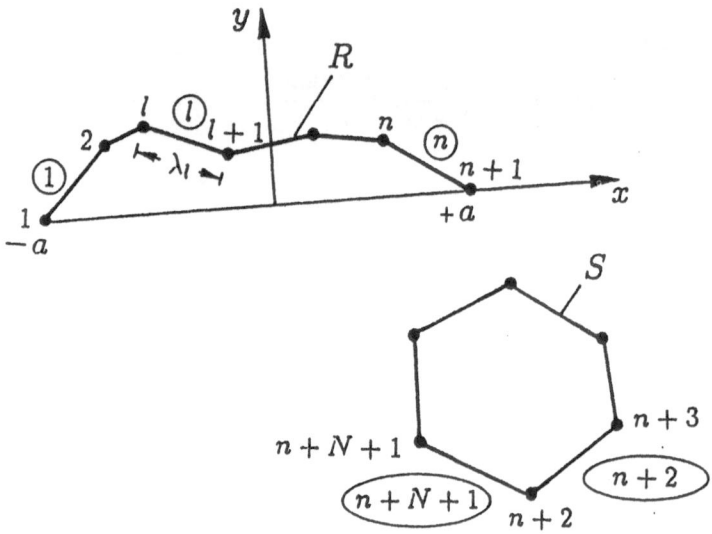

Figure 2: Discretization

The functions $a(t)$ and $g(t)$ along all elements with exception of the crack tip elements are approximated by linear distributions with the unknown complex coefficients A_l. Along the crack tip elements 1 and n the $1/\sqrt{r}$ singular behavior is taken into account, so that e.g. the dislocation distribution is represented by

$$g_l(s) \quad = \quad A_l \frac{1-s}{2} + A_{l+1} \frac{1+s}{2} \quad , \qquad l \ne 1, n, n + N + 1$$

$$g_{n+N+1}(s) \quad = \quad A_{n+N+1} \frac{1-s}{2} + A_{n+2} \frac{1+s}{2} \quad ,$$

$$g_1(s) \quad = \quad A_1 \left[\frac{\sqrt{2}}{\sqrt{1+s}} - 1 \right] + A_2 \quad , \tag{10}$$

$$g_n(s) \quad = \quad A_n + A_{n+1} \left[\frac{\sqrt{2}}{\sqrt{1-s}} - 1 \right] \quad .$$

Using these formulas the integral equations can be solved numerically and from the singular parts of the dislocation distribution in the crack tip elements the mode I and mode II stress intensity factors are determined:

$$K_I^+ + iK_{II}^+ = -2\pi\sqrt{2\pi a\tilde{\lambda}_n}\ \overline{A_{n+1}}\ ,$$
$$K_I^- + iK_{II}^- = 2\pi\sqrt{2\pi a\tilde{\lambda}_1}\ \overline{A_1}\ . \tag{11}$$

The '+' and '−' sign denote the right and the left crack tip respectively.

4   Quasistatic crack propagation

The method described in the foregoing section combined with a suitable crack propagation criterion can be applied to predict crack pathes of thermally loaded cracks. In the following it is assumed that the left crack tip is fixed while the right tip is allowed to propagate quasistatically. As propagation criterion the hypothesis of Erdogan and Sih is used. It states that the crack propagates in the direction α of maximum circumferential stress (Fig. 3).

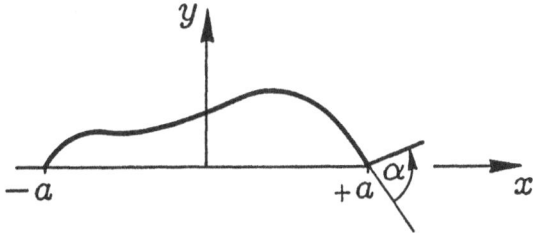

Figure 3: Crack propagation at the right crack tip

This leads to the condition

$$K_I^+ \sin\alpha + K_{II}^+(3\cos\alpha - 1) = 0 \tag{12}$$

which can be approximated for $K_{II}^+ \ll K_I^+$ by

$$\alpha \simeq -\frac{2K_{II}^+}{K_I^+}\ . \tag{13}$$

Having solved the boundary value problem and having calculated the K-factors for the initial problem, a propagation direction is determined by (12). Extending the crack into this direction by a sufficiently small length difference the same procedure is carried out for the new geometry. So step by step the crack path will be determined.

5   Examples

In the following examples the influence of some inhomogeneities or singularities to the crack path is investigated. Starting from an initially straight crack the path of crack propagation is determined. For simplicity it is assumed that the fracture thoughness is zero. As a consequence crack propagation stops if K_I becomes zero. The predicted path of crack propagation is represented by a dashed curve.

5.1 Heat source

The stress field generated by a heat source in an infinite plane is of logarithmic type [1]:

$$\sigma_r \sim \ln r - 1 \quad , \quad \sigma_\varphi \sim \ln r + 1 . \tag{14}$$

Herein σ_r and σ_φ denote the radial and the circumferential stress and r is the distance from the source. All lenght quantities here and in the following are normalized with the initial half crack length.

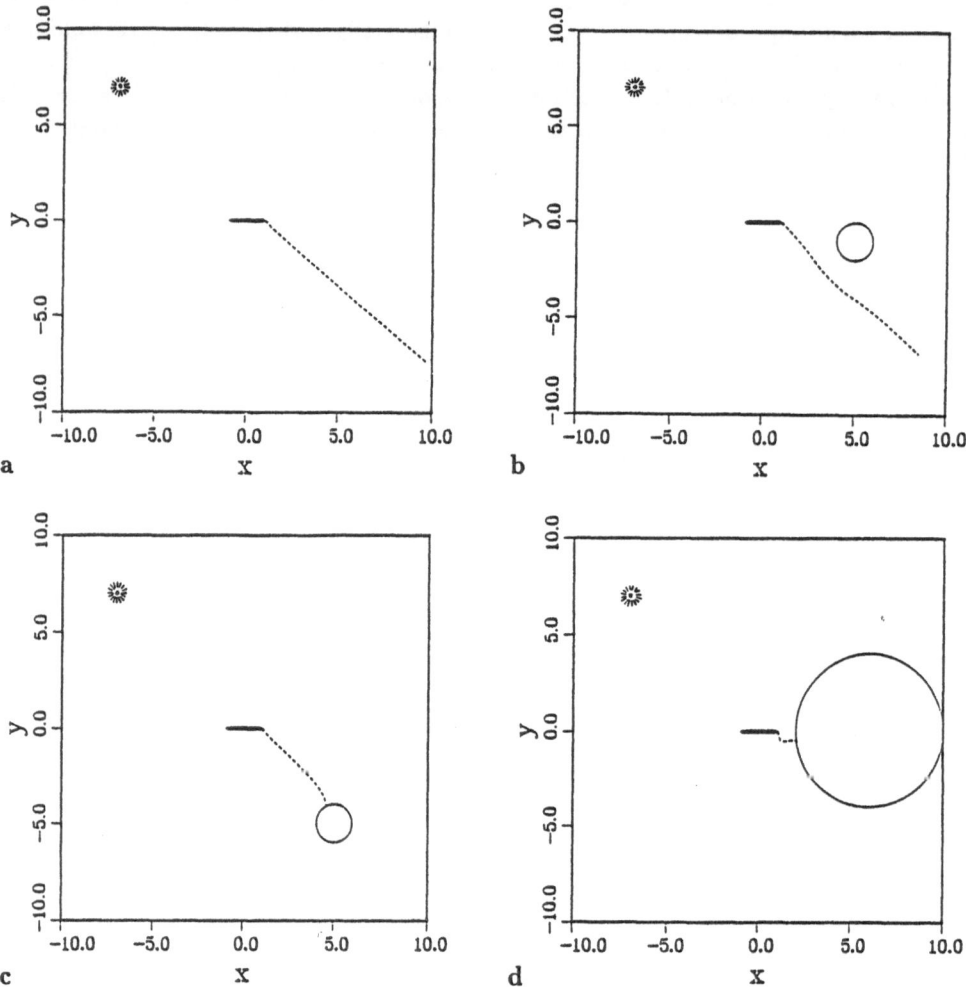

Figure 4: Cracked plane with a heat source and a hole

Figure 4a shows the crack path for an infinite plane with a heat source in a certain distance. If the crack starts to propagate, it immediately turns its direction and moves then nearly straight forward to infinity. The direction of crack propagation is mainly caused by the circumferential stress σ_φ, because $\sigma_\varphi > \sigma_r$.

Figures 4b,c,d illustrate the influence of an additional hole to the crack path. If a small hole is positioned in the neighborhood of the supposed crack path of the undisturbed problem, the crack moves around or into the hole (Fig. 4b,c). If the hole is large enought, the crack changes its first propagation direction strongly and moves directly into the hole (Fig. 4d).

5.2   Negative heat source

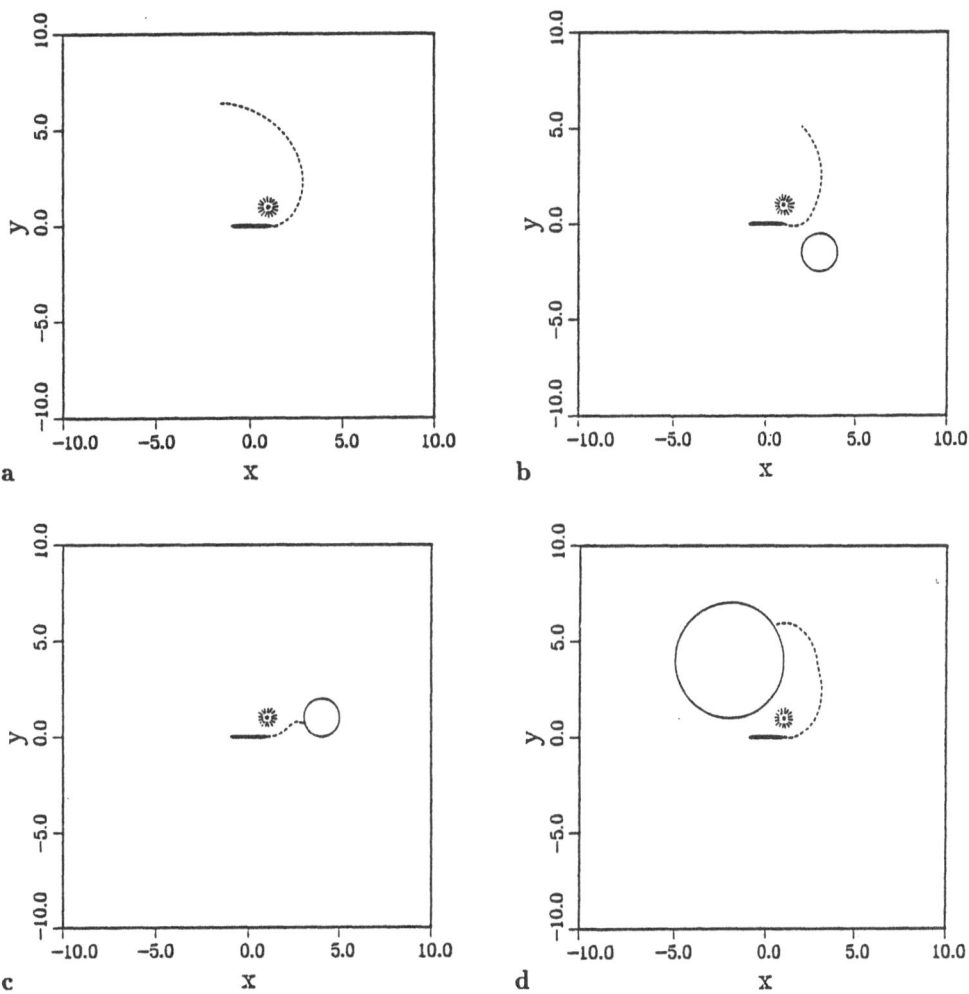

Figure 5: Cracked plane with a negative heat source and a hole

The stress field generated by a negative heat source in an infinite plane is of the same type as (14), but multiplied with -1. As a consequence the behavior of a moving crack is totally changed, compared with that of a positive source. Figure 5a shows the crack path in a plane without a hole, which is now mainly influenced by the stronger radial stresses σ_r. In the plane with an additional hole, the crack path is influenced more or less, dependend on the size and the location of the inhomogeneity (Fig. 5b,c,d).

5.3 Center of dilatation

The stress field generated by a positive center of dilatation in the infinite plane is of the type

$$\sigma_r \sim -\frac{1}{r^2} \quad , \quad \sigma_\varphi \sim \frac{1}{r^2} \tag{15}$$

For a negative center of dilatation the signs in (15) change.

Figures 6a,b show the pathes of a straight initial crack loaded by a positive or negative center of dilatation at a certain position. In the first (second) case the direction of crack propagation is mainly caused by the positive circumferential (radial) stresses.

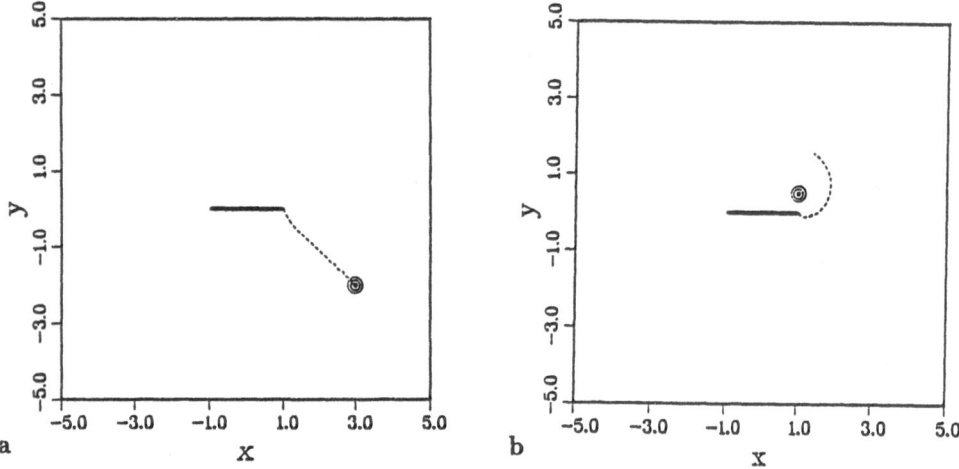

Figure 6: Cracked plane loaded by a) a positive center of dilatation
b) a negative center of dilatation

5.4 Single couple

The stress field induced by a single couple in the infinite plane is as follows

$$\sigma_r = \sigma_\varphi = 0 \quad , \quad \tau_{r\varphi} \sim -\frac{1}{r^2} \tag{16}$$

Figures 9a,b show the crack pathes for a single couple acting counterclockwise or clockwise respectively. The shape of the crack path seems to be similiar as a logarithmic spiral.

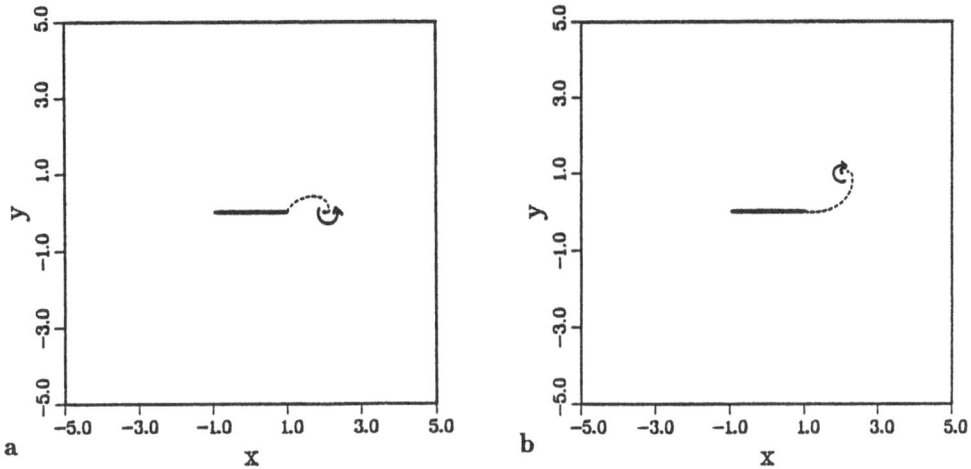

Figure 7: Cracked plane loaded by a single couple

References

[1] Bettin, A.: Beitrag zur Lösung thermo-elastischer Randwertprobleme mit gekrümmten Rissen. Dissertation 1989. Fortschrittberichte VDI, Reihe 18: Mechanik/Bruchmechanik, Nr. 67.

[2] Dreilich, L.: Beiträge zum thermo-elastischen und zum krummlinigen Rissproblem. Darmstadt: Dissertation 1985.

[3] Florence, A. L. und Goodier, J. N.: Thermal stress due to disturbance of uniform heat flow by an insulated ovaloid hole. Trans. ASME, Ser. E. Journal of Applied Mechanics, Vol. 27 (1960), S. 635-639.

[4] Krüger, K.-H.: Die Randintegralmethode zur Lösung ebener Vielrissprobleme der linearen Elastostatik. Darmstadt: Dissertation 1981.

[5] Mußchelischwili, N. I.: Einige Grundaufgaben zur mathematischen Elastizitätstheorie. 5. Aufl.. München: Carl Hanser Verlag 1971.

[6] Sih, G. C.: On the singular character of thermal stresses near a crack tip. Journal of Applied Mechanics (1962), S. 587-589.

[7] Viola, E. und Piva, A.: Crack paths in sheets of brittle materials. Engineering Fracture Mechanics, Vol. 19 (1984) No. 6, S. 1069-1084.

CRACK EXTENSION IN INHOMOGENEOUS MATERIALS

Elmar Steck
Institut für Allgemeine Mechanik und Festigkeitslehre
TU Braunschweig, Gaußstr.14, D-3300 Braunschweig

1 Introduction

Crack extension behaviour in the heat affected zones of welded structures
differs significantly from crack propagation in a homogeneous material. The
crack in a geometrically symmetric precracked specimen under symmetric
loading will deviate from its initial plane, due to changes in the properties of the material caused by the heat treatment during welding.

Mechanical properties in the heat influenced zone vary so strongly, that
they cannot be identified by normal testing procedures. Using the weld simulation technique, however, where large regions of the specimen undergo a
very similar heat treatment, identification of the properties of the material is possible.

A finite element calculation allowing crack extension in arbitrary directions in elastic-plastic bodies, using an extension of the Griffith crack
growth criterion for elastic plastic materials was used to simulate numerically the crack extension behaviour in inhomogeneously weld simulated
specimens.

During quasistatic loading of the specimen, stable crack extension occurs,
and some deviation of the crack extension direction from tho original crack
plane has to be expected. Surprisingly the crack propagates in that region
where the extension of the plastic zone is large and where more energy is
dissipated compared to crack extension in the other directions.

Good agreement was found between experiment and numerical simulation, giving some insight in the energy flow near the crack tip of an inhomogeneous
elastic-plastic material.

Accompanying experimental investigations using grating methods with image
processing confirmed the validity of the results of the FE-calculations.

2 Numerical Simulation of Crack Extension

Numerical simulation of crack growth in elastic-plastic bodies is usually done by analyzing a symmetric specimen under symmetric loading. In this case only half of the body has to be modelled, and the plane containing the crack can be located on the boundary of the model. The basic idea of the most widely used crack growth techniques is to replace the boundary conditions by the reaction forces and then to release these forces in serveral steps [1,2].

This procedure can be readily adopted for non-symmetric problems. However, an automatic re-meshing algorithm has to be implemented to allow crack extension in an arbitrary direction. Details are given in [3].

To calculate the energy release rate, it was necessary to allow finite crack steps in arbitrary directions. A special control strategy allowing testwise opening of all FE-edges adjacent to the crack tip was implemented into the FE-program. The crack extension path which led to the maximum energy release rate could then be chosen for further calculations.

2.1 Energy Flow

Loading a body by forces $\underline{F}_i$, or by displacements $\underline{u}_i$, stresses σ_{kl} and strains ε_{kl} are induced. They give the specific work

$$W_s = \int_0^\varepsilon \sigma_{ij} \, d\varepsilon_{ij} \tag{1}$$

Integrating over the volume of the body, the strain energy

$$W_{ges} = \int_V W_s \, dV \tag{2}$$

can be calculated. If a crack is extended at given forces or displacements, three energy consuming or energy producing processes have to be considered:
1. If the body is loaded by forces $\underline{F}_i$ the points where the forces act will be displaced by $d\underline{u}_i$. The released energy is

$$dW_{load} = \underline{F}_i \cdot d\underline{u}_i \qquad (3)$$

If the body is loaded by given displacements this expression vanishes.

2. Changes in stresses and strains will cause some change dW_{el} of the elastically stored energy. (Notation: If energy is elastically stored, dW_{el} is positive).

3. When forming the new surface or shifting the plastic zone, some energy dW_{diss} will be dissipated. In applications considered in this paper, energy dissipation by plastic deformation largely exceeds all other effects. They will therefore be neglected.

The amount of energy available to extend the crack is given by

$$dW_{rel} = dW_{load} - (dW_{el} + dW_{diss}) \qquad (4)$$

Dividing by the crack length da, we arrive at the energy release rate

$$G = \frac{dW_{rel}}{da} \approx \frac{\delta W_{rel}}{\delta a} \qquad (5)$$

2.2 Fracture Criterion

Based on Eqn.(5) a fracture criterion can be given readily:
Crack extension will occur as soon as the energy release rate G reaches a critical value G_c :

$$G = G_c \qquad (6)$$

The direction of crack extension is given by that direction where G has its maximum.

2.3 Numerical Investigations of Weldsimulated Specimens

In order to permit a comparsion of experimental and numerical results, a series of SEN-specimens ($150 \times 30 \times 15$ mm^3), machined from 20 MnMoNi 5 5 steel with large, inhomogeneously weldsimulated regions was tested (Fig.1).

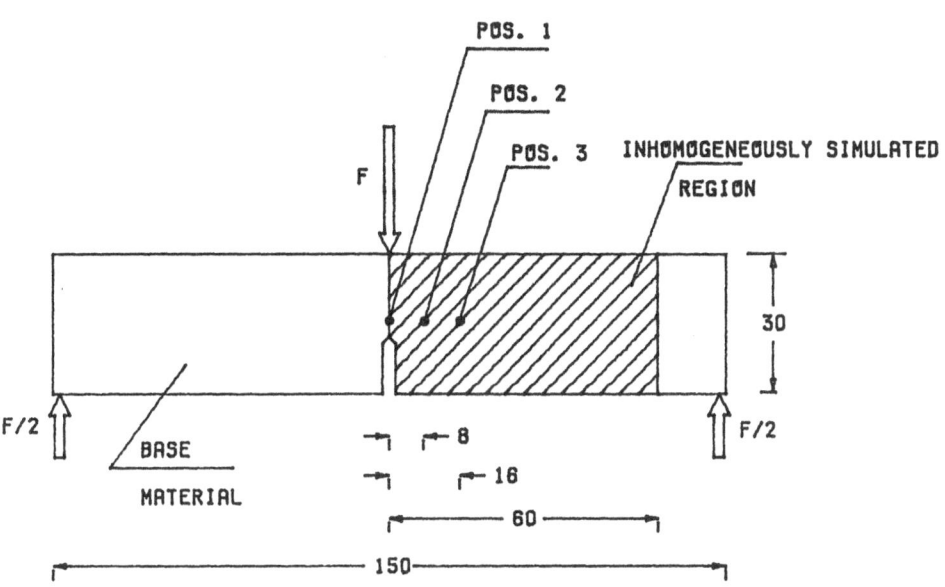

Figure 1: SEN-specimen with inhomogeneously weld simulated area [4]

Microsamples with a diameter of 3 mm were taken from the base material and from the inhomogeneously weldsimulated regions at different positions to give the stress-strain relations (Fig.2). Figure 3 shows the finite element mesh related to the specimens.

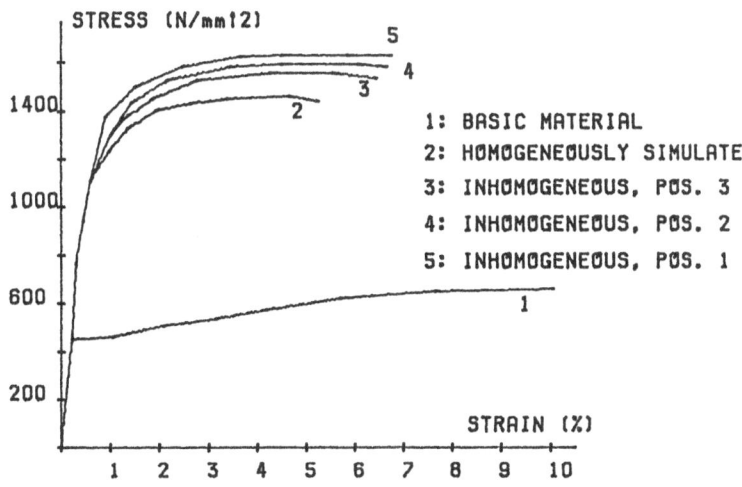

Figure 2: Stress-strain relations of different weld simulated samples [4]

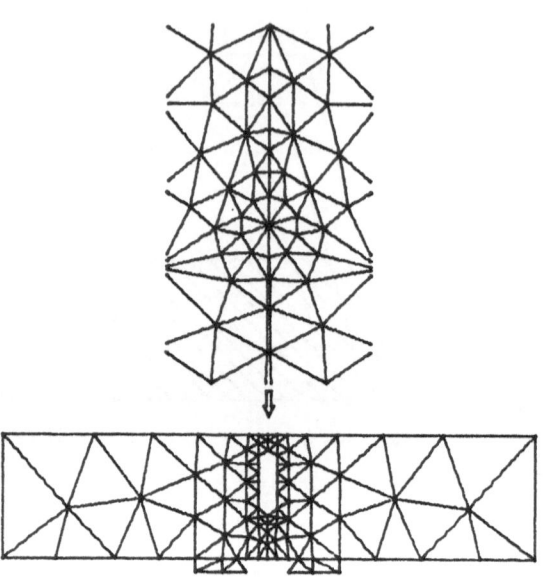

Figure 3: Finite element mesh

Figures 4a and 4b show results of energy flow calculations for a given crack extension step, assuming plane stress and plane strain conditions. The abbreviations at the side of the diagram have the same meaning as in the section on energy flow. As the experiments were displacement controlled, the released energy dW$_{rel}$ equals −dW$_{ges}$.

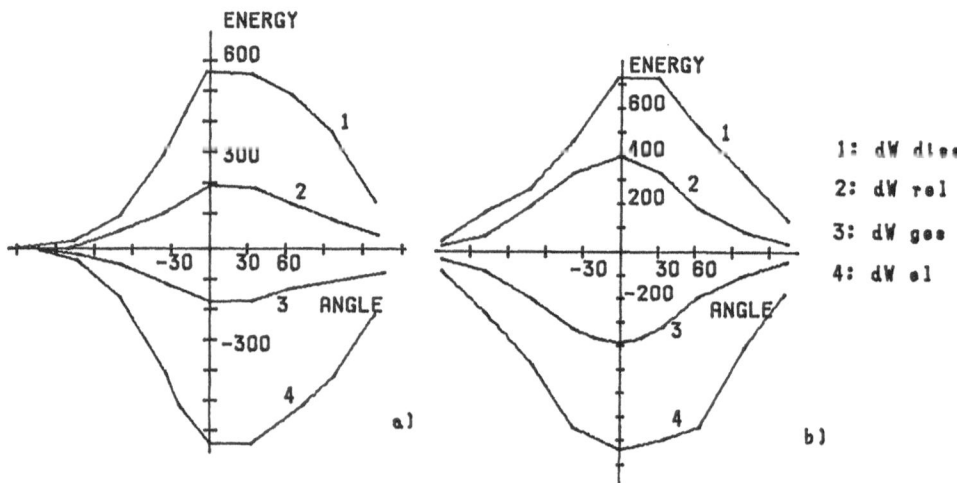

Figure 4: Energy flow, a) plane stress, b) plane strain

During quasistatic loading of the specimen stable crack extension occurs, and some deviation of the crack extension direction from the original crack plane has to be expected. The crack propagates in that region where the extension of the plastic zone is large, and where more energy is dissipated compared to crack extension in other directions.

Examining the diagrams, the effect of crack deviation into the area of a larger plastic zone can be explained. As could be expected the upper curve, representing the dissipated energy, has its maximum at an angle of deviation of about 30°. However, due to the large amount of elastic energy stored in the plastic zone, the maximum of the energy released by elastic unloading lies at about 30°, too. In the case of plane stress the angle of deviation of the maximum energy release rate which was calculated by interpolation was found to be 19.4°. This is in good agreement with the experimental result of about 25° at the surface of the specimen.

Under plane strain conditions, which govern the deformation state in the center of a thick-walled specimen, no significant deviation of the crack is predicted (Fig.4b).

Looking at these results from a more general point of view, it has to be stated that the material properties at some distance of the crack tip have a significant influence on the fracture behaviour. This is of special importance for welded structures, where small brittle zones very often are embedded in large plastic regions. Due to their influence on the energy flow to the crack tip, these plastic zones have to be taken into account when the overall load carrying capacity of a welded structure is evaluated. In many cases this will lead to less conservative results.

For a more detailed examination of crack behaviour and strain- and stress-fields in cracked specimens, experimental methods are under development which allow by the use of grating methods and image processing the measurement of these magnitudes over extended areas of the specimen [5].

3 Grating Methods

The grating methods are based on optical marks, which are combined in a given manner to the considered object surface. The relationship between object, marks, and their image leads to the geometrical properties of its
surface.

Deformation and strain were measured at a plane specimen with a crack
loaded in bending, according to Fig.1. The whole-field measurement was
performed over a range around the tip of a crack, which, growing from the
ground of the notch, extended vertically towards the longitudinal axis of
the specimen with increasing load.

The grating structure was recorded with a camera (Rollei 6006 Reseau) for
different steps of loading or deformation, respectively. Fig.5 shows a grating structure covering the field around the crack. For this case only the
plane deformation was considered.

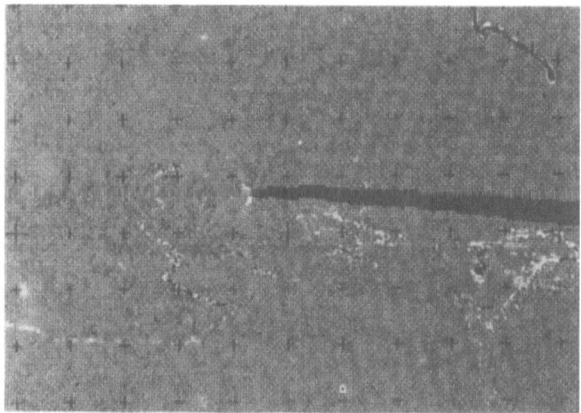

<u>Figure 5:</u> Grating field around the crack in the deformed state of the
object with 10 lines/mm (section), including reseau marks of the
recording camera

4 Image Processing

Calculating strain components from images of a deformed grid requires high accuracy of the related image processing methods.

The film, in consequence, must be digitized in e.g. 10 by 10 segments with a conventional CCD-camera to obtain the inherent resolution. The segmentation is performed using cross marks supplied by a Rollei-Reseau-Camera. These marks are taken onto film with an absolute precision of about 0.5μ together with the deformed grid. The marks are needed to connect the segments over the whole area but, moreover, they are used to correct a possible film deformation due to the developing process.

In order to prevent different sources of errors, the whole image evaluation proceeds in the following steps:

1. Digitizing a high precision quadratic refence grid and evaluating its coordinates.
2. Calculation of a related displacement vector correcting the grid distortion caused by the digitizing system.
3. Calculation and correction of the grating coordiantes in different deformation states.
4. Smoothing and interpolation of missing coordinates.
5. Calculation of displacement fields with respect to the undeformed state.
6. Calculating the plane strain field using a large deformation theory and smoothening.

The coordinates of the deformed grid are determined automatically in a given area. The related program needs serveral initial parameters, namely the pitch of the grating, the thickness and direction of the lines, an index of a starting point etc., which have to be submitted interactively at the beginning of the evaluation. The program then generates a consistent correlation filter and it determines about 10 to 20 points per second on a Microvax II-computer.

These raw data are analysed with a smoothing procedure which detects gross errors, and which interpolates missing points.

From a reference grid, generally assigned to the undeformed state of the surface and from a deformed grid, two displacement matrices are determined, which contain the plastic deformation and a rigid body motion (translation and rotation). The latter can be caused by a large deformation as well as by a displacement of the experimental setup and of the film when being digitized.

As a result three matrices are derived, containing the raw values ε_x, ε_y and ε_{xy} of the strain tensor. These values can be smoothened further over 2 by 2 or 3 by 3 meshes when plotting isolines, if too much noise is present. This, however, reduces the resolution of the strain field.

The described grating method, including image evaluation, was applied to crack propagation in the SEN-specimen given in Fig.1, loaded in bending. Figures 6 and 7 give, as an example, the distribution of the strain component ε_y measured by the whole field method (Fig.6), and calculated by the finite element program (Fig.7).

5 Acknowledgement

This project has been supported by the Deutsche Forschungsgemeinschaft (DFG) under contract Ste 258/4 and Ri 339/7.

References

[1] Andersson, H.: A Finite-Element Representation of Stable Crack Growth. Journal of Mechanics and Physics of Solids, Vol. 27 (1973), p.337.
[2] Du, S. and Lee, J.D.: Variations of Various Fracture Parameters during the Process of Subcritical Crack Growth.Engineering Fracture Mechanics, Vol. 17 (1983), p.173.
[3] Westendorf, H.: Numerische Simulation stabilen Risswachstums in inhomo-genen elastisch-plastischen Materialien. DVS-Verlag, Düsseldorf 1988.
[4] Linnemann, R.: Beitrag zur Bewertung von Schweißnahtfehlern mittels bruchmechanischer Methoden. Diss., TU Braunschweig, 1987.
[5] Andresen, K., Ritter, R. and E. Steck: Theoretical and Experimental Investigations of Fracture by FEM and Grating Methods. Proceedings Europ. Symp. on Elastic-Plastic Fracture Mechanics, Freiburg, Oct. 9-11, 1989

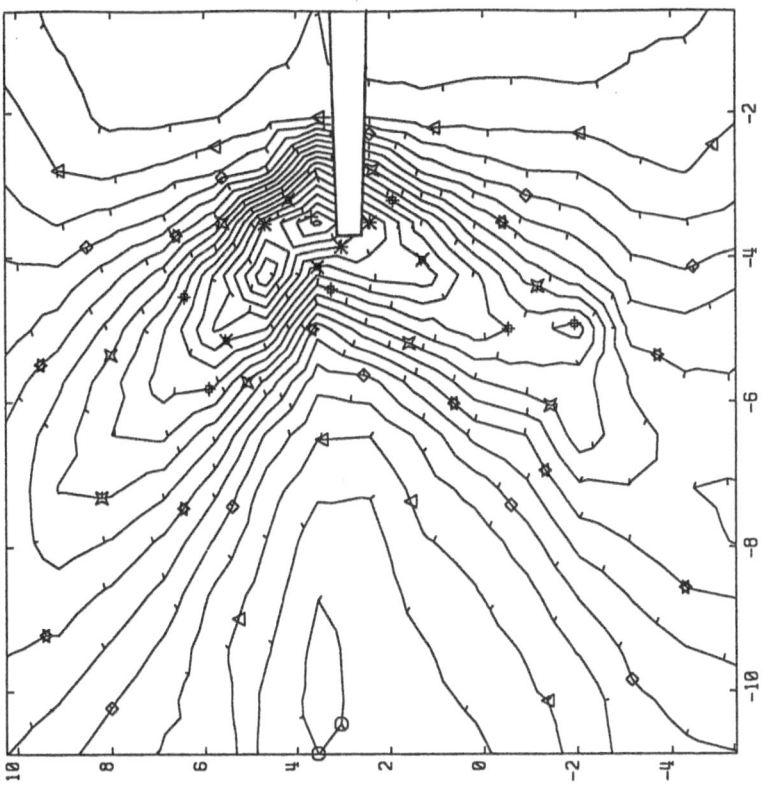

□ : -1.00e-02
○ : 0. e+00
△ : 1.00e-02
◇ : 2.00e-02
✿ : 3.00e-02
¤ : 4.00e-02
✦ : 5.00e-02
✕ : 6.00e-02
✳ : 7.00e-02
+ : 8.00e-02
✕ : 9.00e-02

Figure 6: Strain component ε_y obtained by grating method

104

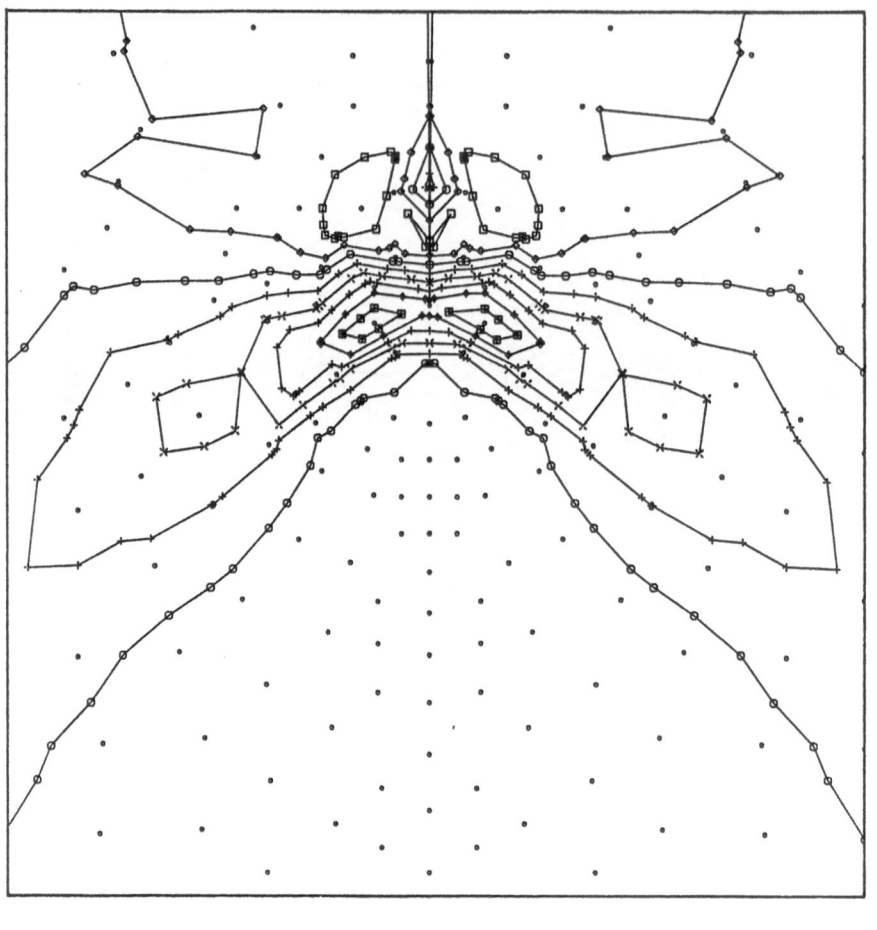

▽	-0.210E-01
▲	-0.140E-01
▢	-0.700E-02
◇	0.000E 00
○	0.700E-02
✧	0.140E-01
✕	0.210E-01
+	0.280E-01
◆	0.350E-01
⊞	0.420E-01

Figure 7: Strain component ε_y calculated by FEM (plane strain)

Thermal Cracking in Fibrous Composites
Using the Normal Stress Ratio

Carl T. Herakovich & J. Mark Duva
Applied Mechanics Program
University of Virginia
Charlottesville, VA 22903, USA

ABSTRACT

The normal stress ratio theory is used to describe crack growth from a notch in a constrained unidirectional graphite-epoxy composite plate. The circular plate is fixed around its circumference and subjected to a uniform temperature change. Stress distributions are obtained from a finite element analysis for circular holes and oval slots of varying size. The predicted crack initiation site and critical temperature drop to initiate crack growth are both functions of the notch geometry and fiber orientation (in the case of a slot). Crack growth is predicted to be parallel to the fiber direction in all cases.

INTRODUCTION

The accurate description of crack growth in fibrous composites is of fundamental importance in understanding the failure of these highly orthotropic, nonhomogeneous materials. Because of orthotropy, crack growth is not typically self-similar. Because of the heterogeneous microstructure, the validity of predictions based on homogeneous material models is uncertain. That is, the stress variations on the scale of the microstructure may be important in predicting failure, and these variations are explicitly excluded from a homogeneous material model.

The results of a theoretical investigation of crack growth in notched unidirectional graphite-epoxy plates constrained against displacement at their boundaries and subjected to a uniform temperature change are presented herein. The composite has been assumed to be homogeneous and orthotropic. The later comparison of these results with experiment will provide a critical test of the normal stress ratio theory (described below) in the context of biaxial loading.

The normal stress ratio was first proposed by Buczek & Herakovich [1] to model direction of crack growth in composites. It has been used to predict the crack initiation site and the critical load for initiation of a crack [2,3,4]. The theory is an extension of the maximum normal stress theory (proposed by Erdogan & Sih [5]) that accounts for the orthotropic strength of unidirectional composites.

Finite element stress analysis using ANSYS [6] were performed to obtain stress distributions under thermal loading. The known stresses were then used in the normal stress ratio theory to predict the characteristics of crack growth from the notch.

PROBLEM FORMULATION

A circular plate of a unidirectional composite containing a centered notch, either a circular hole or an oval slot, is fixed at its periphery (Fig. 1) and subjected to a quasistatic uniform negative temperature change. The composite is modelled as homogeneous and orthotropic whose plane stress thermo-elastic constitutive equation can be expressed in the form:

$$\begin{Bmatrix} \sigma_1 \\ \sigma_2 \\ \sigma_{12} \end{Bmatrix} = \begin{bmatrix} Q_{11} & Q_{12} & 0 \\ Q_{12} & Q_{22} & 0 \\ 0 & 0 & Q_{66} \end{bmatrix} \begin{Bmatrix} \varepsilon_1 - \alpha_1 \Delta T \\ \varepsilon_2 - \alpha_2 \Delta T \\ \gamma_{12} \end{Bmatrix}. \tag{1}$$

For the case of off-axis fibers the constitutive equation takes the form:

$$\begin{Bmatrix} \sigma_x \\ \sigma_y \\ \sigma_{xy} \end{Bmatrix} = \begin{bmatrix} \bar{Q}_{11} & \bar{Q}_{12} & \bar{Q}_{16} \\ \bar{Q}_{12} & \bar{Q}_{22} & \bar{Q}_{26} \\ \bar{Q}_{16} & \bar{Q}_{26} & \bar{Q}_{66} \end{bmatrix} \begin{Bmatrix} \varepsilon_x - \alpha_x \Delta T \\ \varepsilon_y - \alpha_y \Delta T \\ \gamma_{xy} - \alpha_{xy} \Delta T \end{Bmatrix}. \tag{2}$$

The faces of the plate are stress free. Stress distributions for a variety of notch sizes and fiber/slot orientations were obtained for a graphite-epoxy plate whose properties are given in Table 1.

TABLE 1 - Graphite-Epoxy Properties

E_1(msi)	E_2(msi)	G_{12}(msi)	ν_{12}	X_T(ksi)	Y_T(ksi)	α_1($\mu/°F$)	α_2($\mu/°F$)
21.6	1.96	0.83	0.28 ·	210	7.75	-0.43	13.56

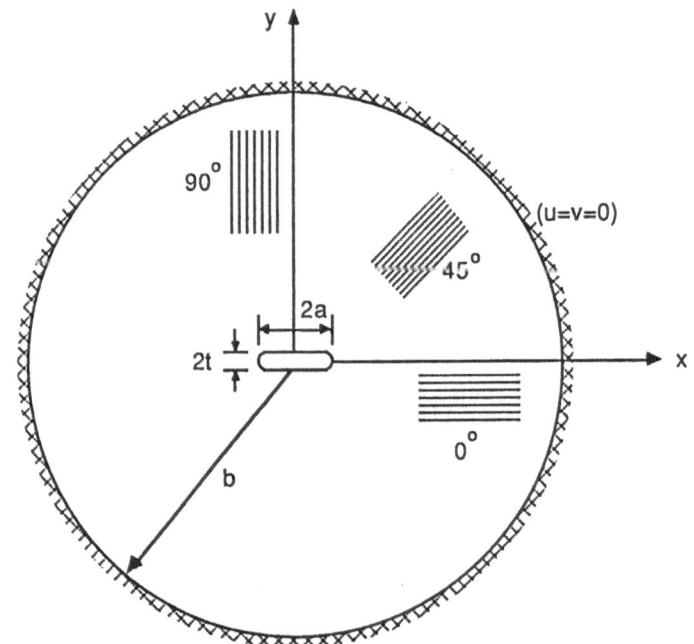

Figure 1. Notched Orthotropic Circular Plate Fixed Along the Edge

THE NORMAL STRESS RATIO

The normal stress ratio (NSR) is a theory for predicting the critical load, initiation site and initial direction of crack growth in fibrous composites. The parameters used in the theory are defined in Figure 2. The theory is based on the assumption that a crack will initiate at any point in an orthotropic material where the ratio of normal stress $\sigma_{\phi\phi}$ to normal tensile strength $T_{\phi\phi}$ at the ϕ plane is unity. Thus, for crack initiation

$$NSR \equiv \frac{\sigma_{\phi\phi}}{T_{\phi\phi}} = 1, \tag{3}$$

where the normal strength $T_{\phi\phi}$ is defined

$$T_{\phi\phi} = X_T\sin^2\beta + Y_T\cos^2\beta. \tag{4}$$

In the above X_T and Y_T are the strength corresponding to failure planes parallel and perpendicular to the fiber direction, respectively. The angle β is the angle between the fibers and the plane of interest. This definition of $T_{\phi\phi}$ has been used because it has not been possible to determine the normal tensile strength experimentally by fracturing a unidirectional composite along an arbitrary plane. The definition does satisfy the three limiting conditions: a) if the materials is isotropic, $T_{\phi\phi}$ is independent of ϕ; b) crack growth on a plane perpendicular to the fibers corresponds to $T_{\phi\phi} = X_T$; c) crack growth along a plane parallel to the fibers corresponds to $T_{\phi\phi} = Y_T$.

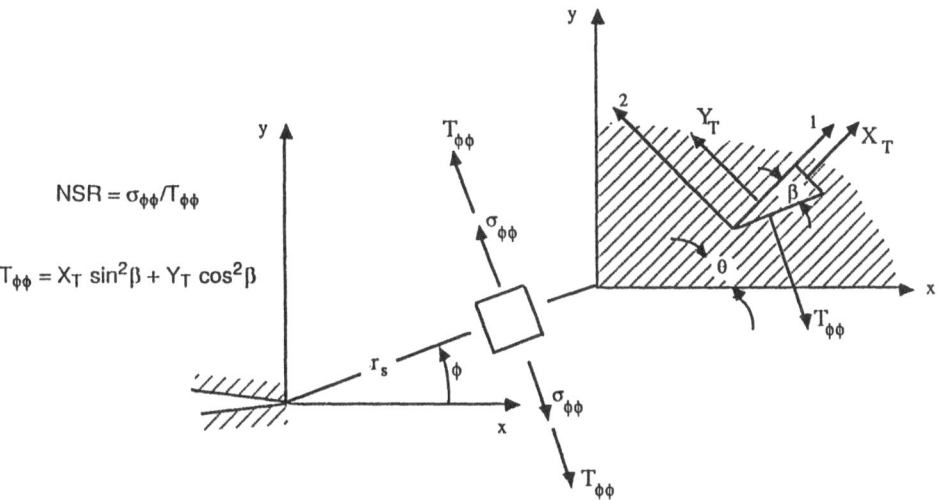

Figure 2. Normal Stress Ratio Parameters

RESULTS AND DISCUSSION

Stress distributions were obtained with ANSYS finite element software. The meshes were composed of 8-node isoparametric quadrilateral elements, with elements concentrated near the boundary of the central notch. Typical meshes are displayed in Fig. 3. Symmetry was exploited to reduce the domain of the analysis to a quarter disk when the possible, otherwise the entire disk was used, resulting in somewhat less mesh refinement in these cases.

In all calculations global equilibrium was well-satisfied and local equilibrium, as judged against the traction free boundary condition at the notch surface, was satisfied to 1 part in several hundred. The calculations were also checked against analytic solutions when possible (isotropic material with a circular notch, orthotropic material with plate of infinite extent). Stresses at selected points on the notch surface delivered by the finite element analysis differed from the analytic solutions by no more than 5%. (The largest discrepancy was for a case in which the analytical solution was also approximate.)

Table 2 is a summary of the cases considered. Material properties for the orthotropic material are as given in Table 1 for graphite-epoxy. For the isotropic material the transverse properties of graphite-epoxy were used. In each instance the outer boundary of the circular plate was constrained against displacement and a uniform temperature change $\Delta T = -100°$ F was imposed.

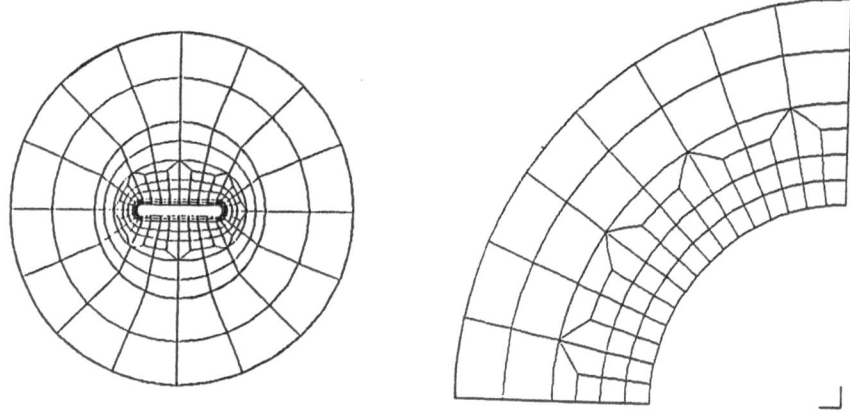

Figure 3. Typical Finite Element Meshes for Fixed Plate

For a circular notch in an isotropic material the constant hoop stress is the only nonzero component at the notch surface due to symmetry. For the orthotropic material there is a compressive stress in the fiber direct due to the negative coefficient of thermal expansion in that direction and the negative temperature change. This provides some degree of relaxation of the tensile stress transverse to the fibers. Fiber orientation is not an issue for the circular notch.

The stresses in the vicinity of a slot are very much a function of the fiber/slot orientation. Figures 4 and 5 show contours for the tensile stress perpendicular to the fibers for parallel and perpendicular fiber/slot orientations, respectively. Both plots are for a/w = 1/16. The maximum stress, shown as the clear region, is relatively low and well away from the notch tip when the slot is perpendicular to the fibers. This geometry is least susceptible to thermal stress induced fracture.

Crack growth predictions are obtained by finding the point on the surface of the notch where the normal stress ratio is largest. Note that the NSR will also be a function or orientation at each position on the notch surface. The critical decrease in temperature is obtained by a linear adjustment of the temperature change such that the maximum NSR=1.0.

Figure 6 shows the critical decrease in temperature as a function of a/w for parallel and perpendicular fiber/slot orientations and for an isotropic material. The trend is opposite to that for mechanical loading in that a larger notch makes the plate less susceptible to fracture. This is because the presence of a large notch diminishes the stress that the constraint produces. As suggested above, the perpendicular orientation is significantly more resistant to cracking than the parallel orientation. The effect of changing the oval slot aspect ratio t/a has not been studied.

TABLE 2 - Cases Analyzed			
Notch Geometry	Materials	a/w	Fiber Orientation
Circular Holes	Orthotropic and Isotropic	1/16, 1/8, 1/4, 1/2	0°
Slots	Orthotropic and Isotropic	1/16, 1/8, 1/4, 1/2	0°, 90°
	Orthotropic	1/4	15°, 22.5°, 45°, 67.5°, 75°

Figure 7 shows the change in the critical decrease in temperature as a function of fiber/slot orientation. The rise to the relatively large value of ΔT is steep, indicating that slot orientations differing much from perpendicular to the fibers will be weak.

The maximum value of the NSR (with respect to orientation of the fracture plane) is plotted as a function of position on the slot surface for 3 fiber orientations in Fig. 8. The spread of these curves is an indication of the degree of scatter one should expect in observations of the fracture initiation site, as a broad curve means the NSR is relatively large over a wide region on the notch surface.

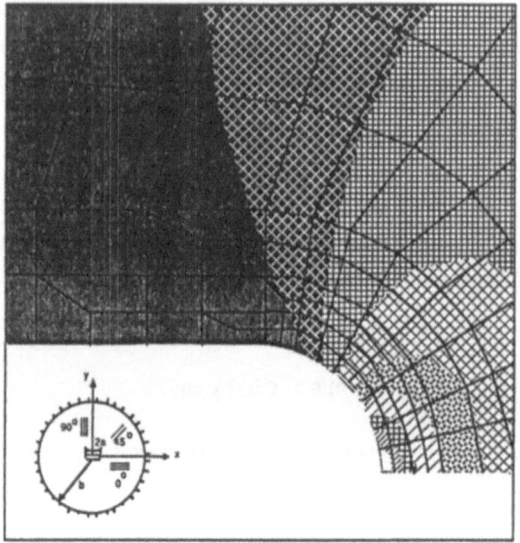

Figure 4. Transverse Stresses in Slotted Plate, θ = 0°

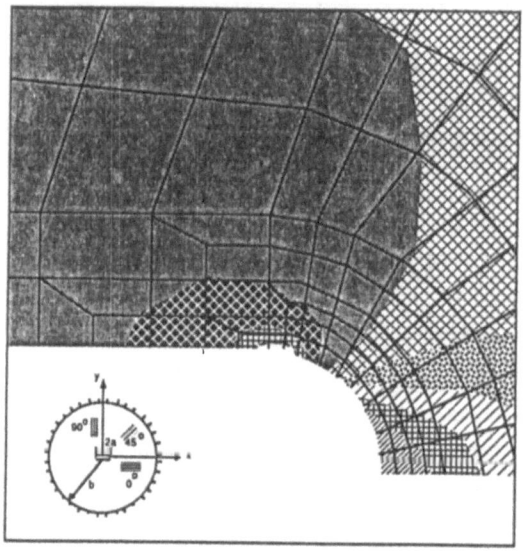

Figure 5. Transverse Stresses in Slotted Plate, θ = 90°

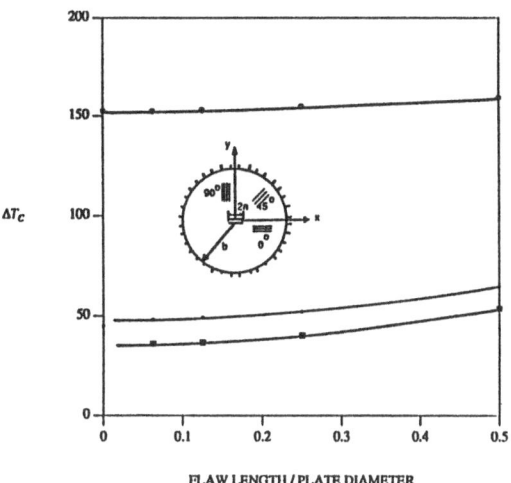

FLAW LENGTH / PLATE DIAMETER

Figure 6. Critical Temperature Change for Crack Initiation

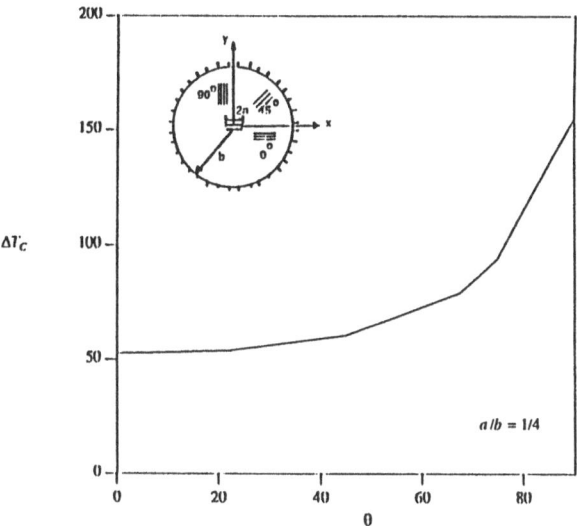

Figure 7. Critical Temperature as a Function of Slot Orientation

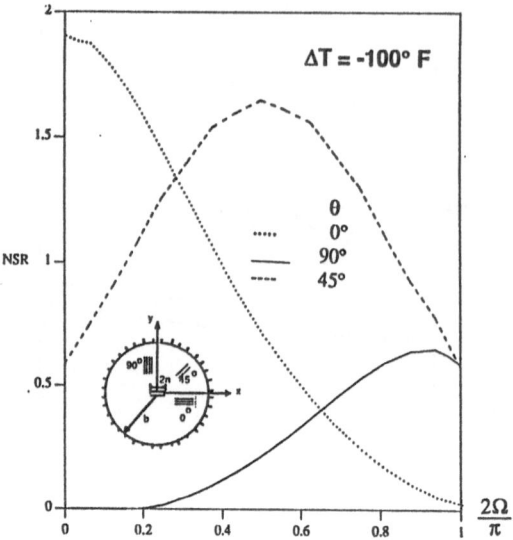

Figure 8. Maximum NSR as a Function of Position on Slot Surface

CONCLUSIONS

Normal stress ratio theory predictions of crack initiation site, (initial) crack propagation direction, and (thermal) initiation load have been calculated for orthotropic materials for a variety of notch geometries. In all cases crack propagation in the fiber direction is predicted although the initiation site is a strong function of fiber/slot orientation, as is the temperature drop required to initiate fracture. The initiation site and the initiation load are also functions of the notch size, though this dependence is weaker.

The most vigorous test of the normal stress ratio theory would be the experimental verification of the prediction of crack initiation site and the initiation load because the dramatic variation in these quantities with fiber/slot orientation and notch size is easily observed. The prediction of the correct propagation direction is a less vigorous test because it varies little in unidirectional composites.

Thermal loading could be a powerful tool for assessing any failure criterion based on a homogeneous material model. The application of mechanical loads equal to the reaction loads in the thermal problem would produce the same failure predictions. Yet in an actual test these different loads would give rise to different stress states on the scale of the composite microstructure. If the variation of the stresses on the scale of the microstructure proves to be important, then *no* homogeneous material fracture criterion can be satisfactorily applied in general.

ACKNOWLEDGEMENT

This work was supported through the Center for Light Thermal Structures which obtains its funding through the Academic Enhancement Program of the University of Virginia.

REFERENCES

1. Buczek, M. B. and Herakovich, C. T., "A Normal Stress Criterion for Crack Extension Direction in Orthotropic Composite Materials", *J. Composite Materials 19*, 544-578 (1985).

2. Beuth, J. L., Jr. and Herakovich, C. T., "On Fracture of Fibrous Composites", *Composites '86: Recent Advances in Japan and the United States*, (Kawata, Umekawa, and Kobayashi, eds.) Prof. Japan-U.S., CCM-III, Tokyo, 267-277 (1986).

3. Gurdal, Z. and Herakovich, C. T., "Effect of Initial Flaw Shape on Crack Extension in Orthotropic Composite Materials", *Theoret. Appl. Fracture Mech. 8*, 59-75 (1987).

4. Beuth J. L., Jr. and Herakovich, C. T., "Analysis of Crack Extension in Anisotropic Materials Based on Local Normal Stress", *Theoret. Appl. Fracture Mech. 11*, 27-46 (1989).

5. Erdogan, F. and Sih, G. C., "On the Crack Extension in Plates Under Plane Loading and Transverse Shear", ASME, *J. Basic Eng. 85*, 519-527 (1963).

6. Swanson Analysis Systems, Inc.,ANSYS, P.O. Box 65, Houston, PA 15342.

THE EFFECT OF A PLASTIC ZONE AROUND A FIBER ON THE FRACTURE RESISTANCE OF A FIBER REINFORCED COMPOSITE

Y.Q. Wang   and   K.P. Herrmann
Laboratorium für Technische Mechanik
Universität Paderborn,   West Germany

ABSTRACT

This study deals with the fracture mechanical investigation of the interaction of plastic zones around fibers in brittle fiber-ductile matrix composites originated by a combination of thermal and mechanical loads with existing microcracks.

INTRODUCTION

The damage growth in composite materials is usually caused by existing micro-cracks [1-3]. The distribution of the microstresses will affect growth of the microcracks and thus also a change of the macromechanical properties [1-8]. Meantimes, the behaviour of microcracks in a composite depends significantly on the properties of the interface between the fibers and the matrix. In this paper, stable growing radial cracks near to fiber-matrix interfaces(C_1 in Fig.1) are studied by means of a unit cell which consists of a ductile matrix and a brittle fiber and is under cooling and tensile loads. Ordinary, the coefficient of linear thermal expansion of the fiber is smaller than that of the matrix for many metal-matrix composites, i.e. $\alpha_f < \alpha_m$. In this situation, compressive radial stresses occur over the fiber-matrix interface and tensile circumferential stresses and axial stresses within the matrix(cf.Fig.2). Therefore, the radial longitudinal cracks are especially dangerous concerning the strength of the composite material. According to investigations performed in the past [9-12], it was found that there often exists a plastic zone around a fibre(cf.Fig.1) because of the stress concentration in the interface between the fibre and the matrix. Fig.2 shows the distributions of the circumferential stresses and the axial stresses in the matrix which decrease rapidly inside of the plastic zone. Therefore, associated with the plastic energy dissipation at the crack tip, the existence of the plastic zone may improve the fracture resistance of fibrous composites. It is possible to sustain a stable crack growth or arrest a crack in the plastic zone. Some experimental and theoretical investigations concerning the stable crack growth of ductile materials have been performed [13-17]. Shih et al used for a numerical assessment Paris' tearing modulus $T_J = (E/\sigma_0{}^2)(dJ/da)$ as a crack extension parameter, where E means Young's modulus and σ_0 is the flow stress. M.Saka

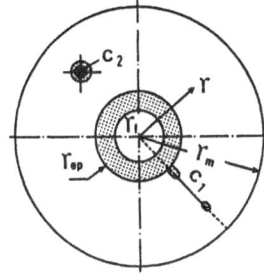

PLASTIC AREA

CRACK AREA

P PLASTIC AREA

E − ELASTIC AREA

Fig.1   A composite unit cell
with two different crack types

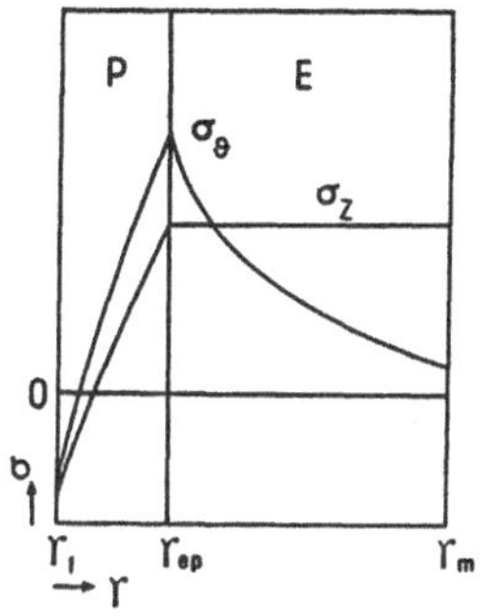

Fig.2   Stress distributions in a partially plastified matrix material

et al [13] presented the rate of the crack tip energy dissipation during crack extension $T_W = (1/R)(E/\sigma_0^2)(dW_p/da)$ as a fracture parameter, where R is the characteristic radius of the intense strain region at the growing crack tip, W_p is the plastic work in the region. F.E.Brust et al [14-15] used a path-independent integral, labeled T^* as a parameter to characterize a stable elastic-plastic crack growth. But for the present problem, there is a fiber-matrix interface located near to the crack tip and the crack tip itself may also be placed within the plastic zone around the fiber. It is not easy to use J, T_J and T^* for a detailed crack analysis. Meantimes the energy dissipation in the plastic zone at the crack tip can not be separated from that in the plastic zone around the fiber when the two plastic zones contact each other. For these reasons, the stable crack growth is described here by means of the fracture mechanical quantities of the energy release rate and the crack tip opening angle. A special finite element program was developed for the numerial calculation of the energy release rate. The release of the nodal forces at that node representing the crack tip was carried out in steps. The profiles of the crack surfaces in initial states and during growing processes are compared and the plastic wake region along the growing crack is also given in the paper. Based on these analyses, the effects of the plastic zone around fibres on the fracture resistance of fibrous composites are discussed.

BASIC FORMULATIONS

1. Constitutive Equations

The analysis is based on the von Mises yield condition associated with the flow rule. The incremental method is applied in the plastic analysis. The matrix is regarded as an elasto-perfectly plastic material. The constitutive equations are:

$$[\sigma] = [D]_E [\epsilon - \epsilon^T] \qquad \text{for } \sigma \leqslant \sigma_y$$

$$[\Delta\sigma] = [D]_{EP}[\Delta\epsilon - \Delta\epsilon^T] \qquad \text{for } \sigma > \sigma_y \qquad (1)$$

where $[D]_E$ is the elastic matrix which is described by a generalized Hooke's law, ϵ^T is the temperature strain, $[D]_{EP}$ is the elasto-plastic matrix. According to the Prandl-Reuss theory $[D]_{EP}$ reads:

$$[D]_{EP} = [D]_E - \frac{[D]_E [\partial\bar{\sigma} / \partial[\sigma]][\partial\sigma / \partial[\sigma]]^T [D]_E}{[\partial\bar{\sigma} / \partial[\sigma]]^T [D]_E [\partial\bar{\sigma} / \partial[\sigma]]} \qquad (2)$$

where $\bar{\sigma}$ is the equivalent stress.

Because the unit cell is very long in comparison to the dimensions of its cross section, the research is based on a generalized plane strain assumption, that means that ϵ_z keeps constant over the cross section.

2. Crack Extension Parameters

2.1 Energy release rate

If a crack extends by a small amount $\Delta_i a$, the stress σ_y along the new extended crack surface will be released and there will be an opening displacement v. For simplicity, this process can be regarded as a continuous process as shown in Fig.3. The crack tip opens when the stress σ_y along $\Delta_i a$ is gradually released. Obviously, the released energy is the volume covered by the σ_y stress surface and the energy release rate can be expressed as

$$G = \frac{2}{\Delta_i a} \int_0^{\Delta_i a} dx \int_0^{v(0,x)} \sigma_y(x,v) \, dv \qquad (3)$$

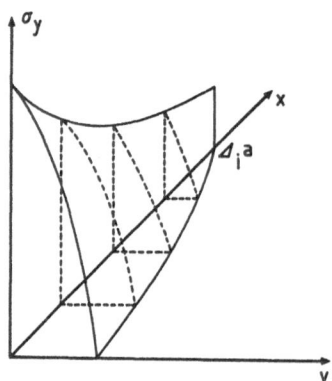

Fig.3 Energy release

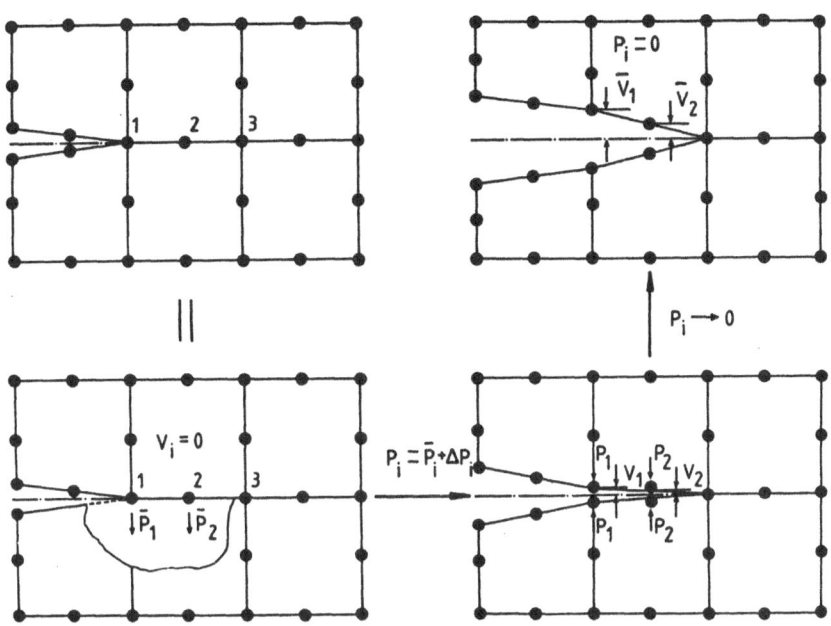

Fig.4   FEM-model for an energy releasing process

The process can be modeled by the FEM-model shown in Fig.4. The nodal forces at the nodes 1 and 2 are released in steps and the energy release rate G can be obtained by means of the following integral:

$$G = \frac{2}{\Delta_i a} \left(\int_0^{v_1} \bar{P}_1 \, dv_1 + \int_0^{v_2} P_2 \, dv_2 \right) \tag{4}$$

2.2 The crack tip opening angle

The crack tip opening angle can be regarded as an alternative fracture parameter describing stable crack growth. In the present analysis, $\Delta_i \delta$ is based on the opening displacement at the first node at a fixed distance, $\Delta_i a$, behind the current crack tip:

$$\Delta_i a = 1.5 \% \, a \tag{5}$$

where   a   is the crack length.

NUMERICAL ANALYSIS OF UNIT CELLS WITH CRACKS

Fig.5 shows a finite element model for the numerical analysis of a cracked composite unit cell. Because of the symmetry of the cell, only one half of the unit cell was considered. The associated boundary conditions read:

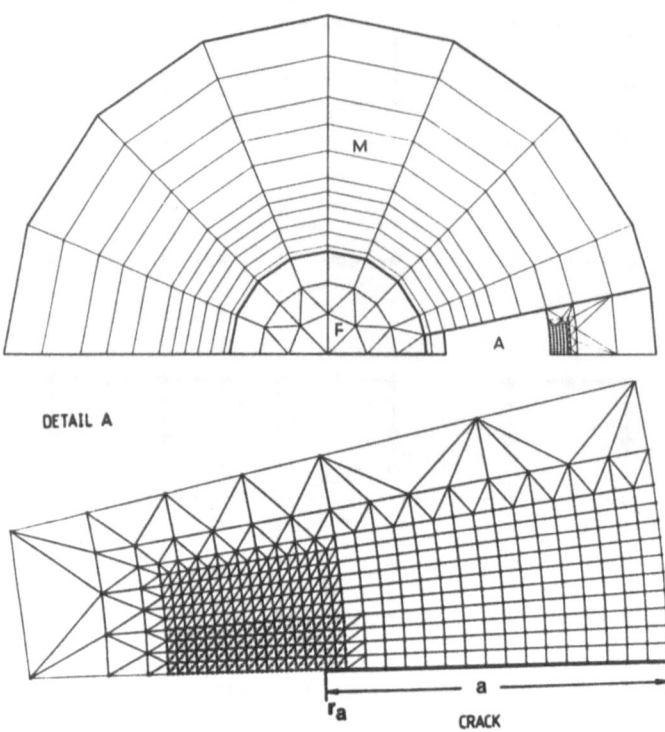

DETAIL A

Fig.5 The finite element model of a cracked unit cell

$$\sigma_r \Big|_{r=r_m} = 0$$

$$\sigma_y \Big|_{y=0} = 0 \qquad \text{for the crack surface} \tag{6}$$

$$u_y \Big|_{y=0} = 0 \qquad \text{for the ligament}$$

For the sake of convenience of data preparation, this unit cell was divided into several substructures. The thick lines in Fig.5 represent the boundaries of the substructures. The crack length is a and the crack propagates towards the fibre. The largest crack propagation length is $\Delta a = 1/3$ a. The stable crack growth with four different positions of the crack tip under at least four different thermal loads ($\Delta T = -100°K, -200°K, -300°K, -400°K$) was calculated. The geometrical parameters and the material properties are

Fiber : $r_f = 3mm$, $\alpha_f = 6.75 \cdot 10^{-6} K^{-1}$, $\gamma_f = 0.21$, $E_f = 8.4 \cdot 10^4 N/mm^2$

Matrix: $r_m = 10mm$, $\alpha_m = 2.39 \cdot 10^{-5} K^{-1}$, $\gamma_m = 0.34$, $E_m = 7.2 \cdot 10^4 N/mm^2$

The positions of the crack tip near to a fiber(cf.Fig.5) are given by $r_a = 3.3, 4.3, 4.7,$ and 5.1mm respectively. The crack length a (cf.Fig.5) is 1.5mm.

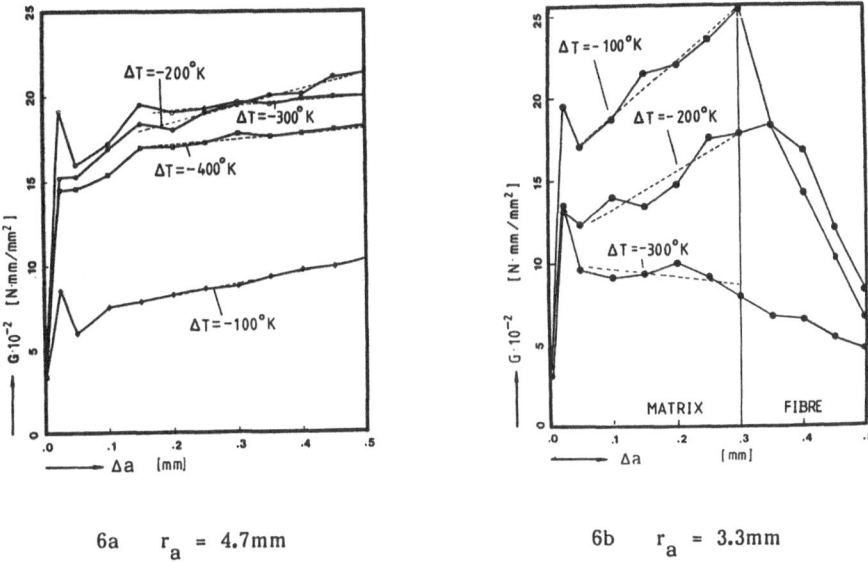

6a r_a = 4.7mm   6b r_a = 3.3mm

Fig.6 Energy release rate G under different thermal loads

Fig.6 shows the energy release rate for r_a = 4.7mm and r_a = 3.3mm, respectively. It can be seen from Fig.6a that the energy release rate increases firstly with increasing thermal load until ΔT = -200°K and decreases gradually for higher thermal load. This phenomenon can be explained by the wake of the plastic zone at the crack tip shown in Fig.7   which gives the plastic area in region A (see Fig.5). For ΔT = -100°K   both plastic zones around the fiber and around the crack tip are small and they don't contact each other. The circumferential stresses near to the crack tip will increase with increasing thermal load and then the energy release rate will increase too. But for ΔT = -200°K both plastic zones start to contact and get together gradually. Because the circumferential stresses decrease rapidly in the plastic zone around the fiber (cf.Fig.2), the stresses near to the crack tip are improved and the energy release rate will not increase for the higher load. This can also be seen clearly from Fig.8 in which the dotted lines represent that thermal load at which the two plastic zones contact. Fig.6b shows energy release rates for a crack tip very near to a fiber where both plastic zones join when the thermal load is rather small. This picture shows the situation when the crack penetrates into the fiber from the matrix side and the energy release rate will decrease very quickly after this penetration.

Fig.9 gives the profiles of a stable crack extension. The crack becomes much sharper after the onset of growth than it was in the initial state. Therefore, the energy release rates are always much smaller at the initial crack growth than those for further stable crack growth.

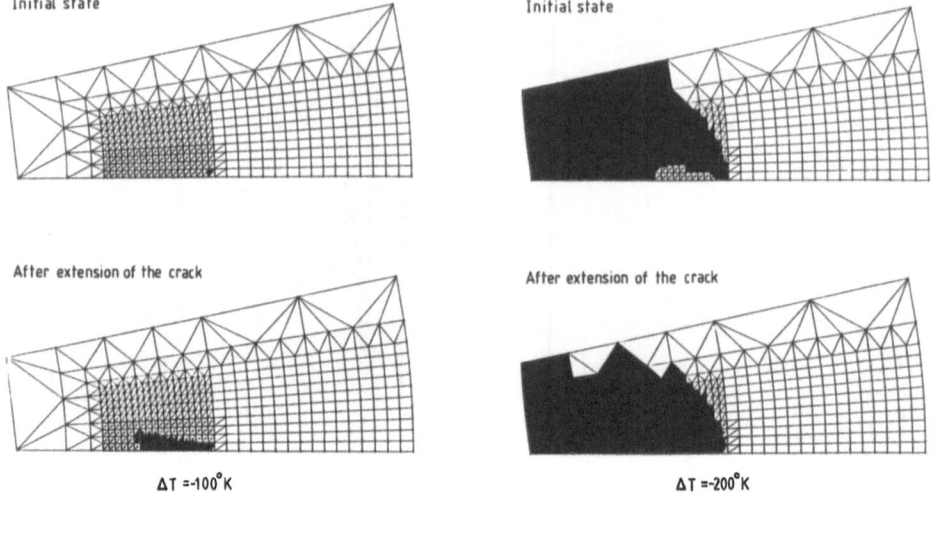

Initial state
After extension of the crack
ΔT =-100°K
Initial state
After extension of the crack
ΔT =-200°K
- Plastic zone

Fig.7 The wakes of the plastic zones for stable crack growth

r_a = 5.1 mm
r_a = 4.7 mm
r_a = 4.3 mm
r_a = 3.3 mm
G 10^-2 [N·mm /mm^2]
0
5
10
15
20
25
0°
-100°
-200°
-300°
-400°
ΔT [K]

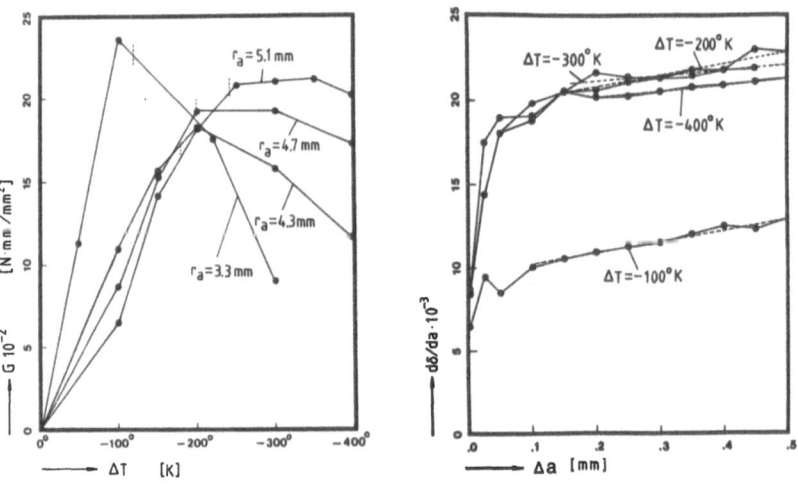

Fig.8 Energy release rate in dependence on temperature

ΔT=-200°K
ΔT=-300°K
ΔT=-400°K
ΔT=-100°K
dδ/da·10^-3
0
5
10
15
20
25
.0
.1
.2
.3
.4
.5
Δa [mm]

Fig.10 Crack tip opening angle for r_a = 4.7mm

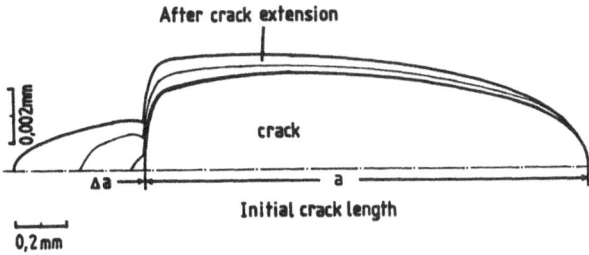

After crack extension

crack

Δa

a

Initial crack length

0,2 mm

0,002 mm

Fig. 9   The profiles of a stable crack extension

Fig.10 shows the curves of the crack tip opening angle which have a certain similarity with the graphs for the energy release rates from Fig.6a.

DISCUSSIONS

The following conclusions can be drawn from the foregoing investigations:

1. The energy release rates at the tips of stable extending cracks near to fibers take maximum values when the plastic zones around the fibers come in contact with the crack tip. Then, the existence of a plastic zone around the fiber will prevent that the energy release rate at the crack tip increases continuously with increasing thermal load.

2. In the present study, the fracture parameter CTOA shows similar graphs in comparison with the curves for the energy release rate at the tip of a thermal crack.

3. Similar to the distribution of the circumferential stresses, the axial stresses decrease rapidly in the plastic zone around the fibers(cf.Fig.2). Therefore, the existence of a plastic zone may also improve the fracture resistance to penny-shaped cracks located in unit cells which are submitted to tensile loads. Because the plastic zones around fibers are mainly caused by thermal loads, it   means that the fracture resistance to penny-shaped cracks of a unit cell under both tensile and thermal loads may be better than that of a cell under tensile load only. Meantimes, because an acting tensile load makes the plastic zone of a cell under thermal load larger, the fracture resistance to radial cracks located inside of unit cells submitted to a combination of thermal and tensile loads may be better than that of unit cells under thermal load only. A detailed investigation is presently on the way.

ACKNOWLEDGEMENT

The support of the Alexander von Humboldt-Foundation for one of the authors (Y.Q.Wang) is gratefully acknowledged.

REFERENCES

1. R.J. Nuismer and S.C. Tan, "The role of matrix cracking in the continuum constitutive behaviour of a damaged composite ply", In Zvi Hashin and Carl T. Herakovich(Eds.), Mechanics of Composite Materials: Recent Advances,Pergamon Press, 437-448(1982).

2. G.P. Sendeckyj, G.E. Maddux and N.A. Tracy, "Comparison of holographic, radiographic and ultrasonic techniques for damage defection in composite materials", Proceedings of 2nd International Conference on Composite Materials, Toronto, 1037-1056(1978).

3. E.F. Olster and R.C. Jones, "Effect of interface on fracture", in Arthur G. Metcalfe(Ed.), Interfaces in Metal Composites, Academic Press,245-284(1974).

4. G.J. Dvorak and M.S.M. Rao, "Thermal stresses in heat-treated fibrous composites", Journal of Applied Mechanics 43, 619-624(1976).

5. R.L. McCullough, "Influence of microstructure on the thermoelastic and transport properties of particulate and short fiber composites", in Zvi Hashin and Carl T. Herakovich(Eds.), Mechanics of Composite Materials: Recent Advances,Pergamon Press, 17-29(1982).

6. R.M. Christensen, "Mechanical Properties of Composite Materials", in Zvi Hashin and Carl T. Herakovich(Eds.), Mechanics of Composite Materials: Recent Advances, Pergamon Press, 1-16(1982).

7. J.D. Achenbach and H. Zhu, "Effect of interfacial zone on mechanical behaviour and failure of fiber reinforced composites", Journal of the Mechanics and Physics of Solids 37, 381-393(1989).

8. B.W. Rosen, S.V. Kulkarni and P.V. McLaughlin, Jr., "Failure and fatigue mechanisms in composite materials", in Carl T. Herakovich(Ed.), Inelastic Behaviour of Composite Materials, AMD-13, ASME, 17-72(1975).

9. K.Herrmann and F. Ferber, "Numerical and experimental investigations of branched thermal crack systems in self-stressed models of unidirectionally reinforced fibrous composites", in S.N. Atluri(Ed.), Computational Mechanics 88, Springer, 3.V.1-8.V.4(1988).

10. K.P. Herrmann and I.M. Mihovsky, "Approximate analytical investigation of the elastic-plastic behaviour of fibrous composites", XVII International Congress of Theoretical and Applied Mechanics, Grenoble, France(1988).

11. K. Herrmann and P. Pawliska, "Finite element analysis of thermal crack growth in self-stressed fiber reinforced composites with partially plastified matrices", International Journal of Fracture 31, R11-R16(1986).

12. K.P. Herrmann and Y.Q. Wang, "Crack analysis for a unit cell of a fiber reinforced composite under axial and thermal loading", ZAMM 70, T304-306(1990).

13. M. Saka, T. Shoji, H. Takahashi and H. Abe, "A criterion based on crack-tip energy dissipation in plane-strain crack growth under large scale yielding", in C.F. Shin and J.P. Gudas(Eds.), Inelastic Crack Analysis, ASTM STP-803, 1130-1158(1983).

14. F.W. Brust, T. Nishioka, S.N. Atluri and M. Nakagari, "Further studies on elastic plastic fracture utilizing the T^* integral", Engineering Fracture Mechanics 22, 1079-1103(1985).

15. F.W. Brust, J.J. McGowan and S.N. Atluri, " A combined numerical/experimental study of ductile crack growth after a large unloading, using T^*, J and CTOA criteria", Engineering Fracture Mechanics 23, 537-550(1986).

16. M.P. Wnuk, "Discontinuous extension of fracture in elastic-plastic deformation field", in C.F.Shin and J.P. Gudas(Eds.), Inelastic Crack Analysis, ASTM, STP 803,1159-1175(1983).

17. S.G. Russell, "An investigation of the fracture process zone near the tip of a steadily propagating tensile crack", International Journal of Solids and Structures 25, 1157-1175(1989).

FAILURE MECHANISMS AND FRACTURE TOUGHNESS OF
DIFFERENT POLYMER COMPOSITE SYSTEMS

K. FRIEDRICH
Polymer & Composites Group,
Technical University Hamburg-Harburg
2100 Hamburg, 90, West-Germany

ABSTRACT

The fracture properties of engineering polymers and composites are strongly affected by two major areas of influence. The first one covers the microstructural parameters of the materials, whereas the second one includes the external testing conditions. In this contribution, it is mainly outlined how area one can determine the fracture characteristics. An introductory section illustrates the variety of microstructural details, for example molecular structure and semicrystalline polymer morphology, and filler related factors, such as volume fraction of reinforcing fibers, their orientation etc. In the following part, effects of these parameters on fracture mechanical properties are discussed. It is distinguished between the fracture behavior of unfilled engineering polymers, of short fiber reinforced, injection molded thermoplastics, and of continuous fiber composite laminates. In the latter group, special emphasis is given to the effect of new, high temperature resistant thermoplastic matrices, for instance PEEK, on the interlaminar fracture energy of the composites.

INTRODUCTION

If materials in use as structural components are subjected to high mechanical loadings, besides the demands for high stiffness and strength very often high values of their fracture toughness are required. This is not only valid for the very large group of metallic materials but also for ceramics and engineering polymers as well as their composites. In all these cases the mechanical properties and the failure behavior are strongly influenced by the microstructural parameters of the particular material. In this contribution the correlation between microstructure and fracture toughness of engineering polymers and their fiber reinforced composites will be outlined in more detail.

MICROSTRUCTURE

Engineering Polymers

With respect to the basic structue of engineering polymers, it must be distinguished between melt-formable thermoplastics and thermosetting resins. Because of the chemical crosslinking of their molecules, the polymers in the latter group are usually stiffer, harder and more brittle than the thermoplastics. Here, an uncrosslinked state predominates in which the molecules are either randomly arranged (amorphous) or partly ordered next to each other. In this semicrystalline condition a morphological structure is built up which very often consists of fine or coarse spherulites [1]. Another important parameter is the molecular weight, i.e. a quantity characterizing the length of the polymer molecules. The longer the molecule chains are, the higher is the molecular weight M_w, which often results in a lower degree of crystallinity and an increase in the toughness of the material.

Additional improvements of the toughness of brittle polymer matrices can be achieved by finely dispersed, tougher particles of a second polymer phase. These particles can cause enlarged zones of local deformation near points of enhanced stress concentration, for example crack tips, thus resulting in a higher amount of energy absorption prior to final breakdown (Fig. 1) [2].

Under the term "engineering polymers" all those polymers are summarized which have a much higher mechanical property profile than conventional plastics like PE, PP, PVC or PS. While the latter are processed in large quantities for minor quality parts, engineering polymers such as PA 6.6 and PC are used as technical components for which high durability and extraordinary mechanical performance are required. Table 1 lists a variety of new "engineering thermoplastics" with especially high temperature resistance. In the other group of "engineering thermosets" used as matrices for high performance composites, epoxy and polyimide resins are of greatest importance.

Polymer Composites

The addition of harder and ultra-high strength components to a polymeric matrix leads to a composite. The objective is to combine the advantageous properties of the single components in a new material.

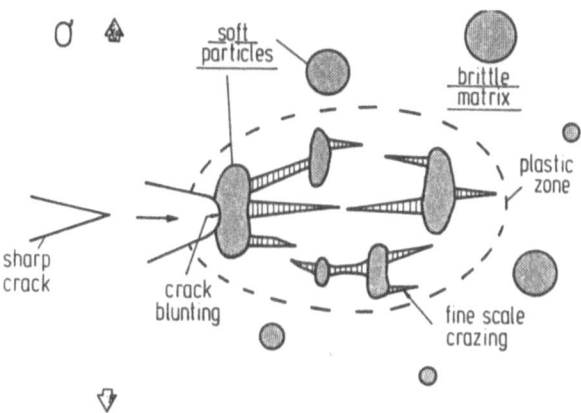

Fig. 1: Mechanisms of energy absorption by tough particles in a brittle polymer matrix in front of a propagating crack

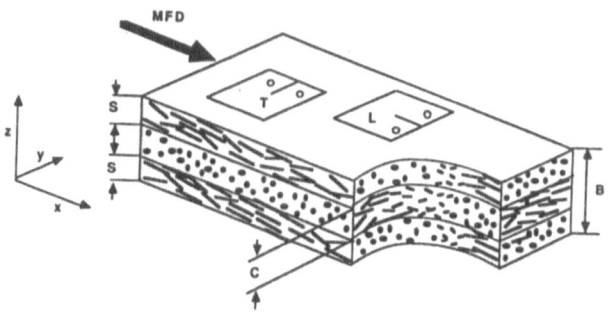

Fig. 2: Schematic of the layered structure in injection molded short fiber reinforced thermoplastic plates

Nr.	Name	Chemical Structure	T_g (°C)	T_m (°C)	E (GPa)$_{flex}$	σ_B (MPa)	ε_B (%)
1	PES		225	—	2.6	84	60
2	PPS		200	300	4.2	75	3
3	PEI		217	—	3.3	105	60
4	PIS		275	—	5.0	63	1.3
5	PEEK		134	335	3.8	100	100
6	LCP		190	280	9.2	196	4

Table 1: List of new, high temperature resistant "engineering thermoplastics"

For instance, reinforced thermoplastics with short glass, carbon or aramid fibers can still be processed into complicated shapes by a relatively simple injection molding technique (advantage of processibility maintained from the thermoplastic matrix). At the same time the fibers yield enhanced stiffness, strength and dimensional stability relative to the unreinforced plastic. During the injection molding process, the fibers are oriented in a special way, and the degree of fiber orientation as well as the orientation distribution depend on the flow conditions in the mold and fiber related parameters such as fiber length and volume fraction. Fundamental studies on the fracture behavior of these structures are often carried out with injection molded plates, from which test samples with defined microstructure can be machined (Fig. 2) [3].

Continuous fiber reinforced polymer systems normally possess a laminate structure consisting of individual layers of unidirectional or woven fibers, with different angles between the individual layers [4]. Fig. 3 illustrates possible, three dimensional lay-ups of such laminates, being built up of individual lamina of continuous fibers embedded in a polymer resin matrix [5]. A lot of other arrangements are possible and can lead to various property profiles in different loading directions.

FRACTURE TOUGHNESS

Fracture Toughness of Engineering Polymers

Testing of the fractue toughness of engineering polymers can in principle be performed following the same guide-lines given for metallic materials in the American standards ASTM/E/399. However, as polymers exhibit normally a much higher ductility, not always all the requirements for the acceptance of the measured stress intensity factors as real fracture toughness values, K_{IC} (in the sense of a plane strain condition value) are fulfilled. Therefore, fracture toughness values of many engineering polymers are only valid for certain material thicknesses and testing conditions (designated as K_C or K_Q values). In these cases the data can only be considered as a basis of comparison for the effects of certain microstructural parameters on the toughness profile of these materials.

If the plasticity in the crack-ground is so high that no crack instability can be achieved the K-concept according to ASTM/E/399

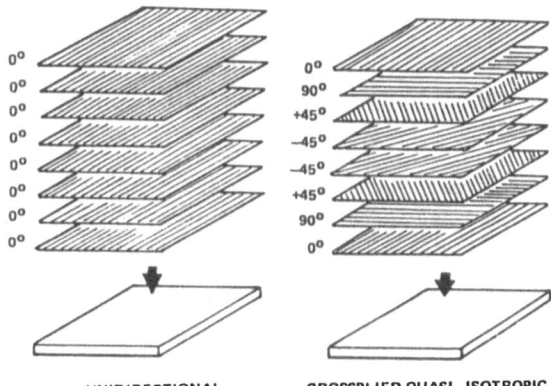

UNIDIRECTIONAL   CROSSPLIED QUASI-ISOTROPIC

Fig. 3: Three-dimensional illustration of possible cross-sections of continuous fiber/polymer matrix composites:
(a) unidirectional lay up
(b) quasi-isotropic lay up [5]

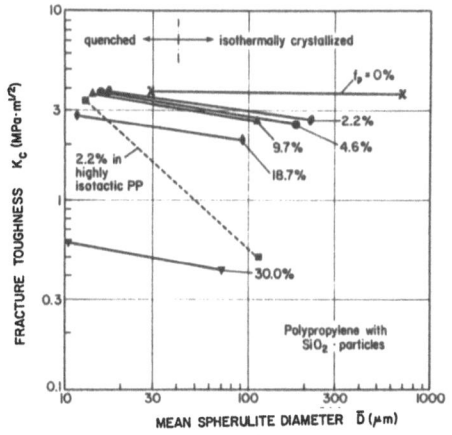

Fig. 4: Fracture toughness K_C as a function of spherulite diameter D and vol.% of SiO_2 particles in isotactic polypropylene [9]

Trade name	Polymer	Toughness K_C |MPa·m$^{1/2}$|
Ultrason E (BASF)	PES	1.0 - 1.4
Liquid Crystal Pol. (Celanese)	LCP	4.0 - 4.8
Victrex 450 G (ICI)	PEEK	5.6 - 7.4
Ryton (Phillips)	PPS	1.1 - 1.3
Delrin 500 (Du Pont)	POM	3.9 - 4.5

Table 2: Ranges of fracture toughness, K_C, of various thermoplastics, as measured at room temperature and relatively low crack opening velocity ($v = 10^0$-10^1 mm/min) [14-17]

should be avoided completely. Under these circumstances methods of the plastic fracture mechanics should be applied. Among these methods the J-integral approach (ASTM/E/813) has got the most frequent attention in recent years [6-8].

Figure 4 gives an impression on the influence of different morphological features on the fracture toughness, K_C, of moderately isotactic polypropylene [9-11]. The K_C value decreases with D and with the embrittlement of the spherulite boundaries due to different amounts of powder-like SiO_2 impurities. Simultaneusly, the fracture behavior changes from a craze domintated, in-plane ductile fracture to a brittle interspherulitic fracture. The breakdown mechanisms and the resulting fracture properties can also be varied in a wide range by the molecular weight and the content of atactic polypropylene [12, 13].

For tougher polymers such as amorphous polycarbonate (PC) testing of the toughness profile by the J-integral method is often more appropriate than the application of the K-concept. Figure 5 shows a corresponding R-curve for this material, and the figure caption contain the J_{IC}-(fracture energy)-value derived from these curves.

For general information, a number of fracture toughness values of numerous newly developed materials is listed in Table 2 [14-17]. It must be mentioned here, that the given data are average values which have been obtained at room temperature and moderate crack opening velocities (1-10 mm/min). Variations of these external parameters can result in drastic effects on the fracture toughness of the materials, which has to be considered in the selection of polymeric materials for different purposes. Further fracture toughness values of other polymers can be found in the books by Kausch [18], Kinloch and Young [19], Williams [20] and Friedrich [21].

Fracture Toughness of Short Fiber Reinforced Engineering Polymers

In this material group, the fracture toughness at a given elastic modulus of the material depends strongly on the sum of the individual energy absorbing mechanisms in the damage zone in front of the crack (Fig. 6): a) fiber fracture, b) fiber/matrix separation, c) fiber pull-out and d) deformation and fracture of the polymer matrix.

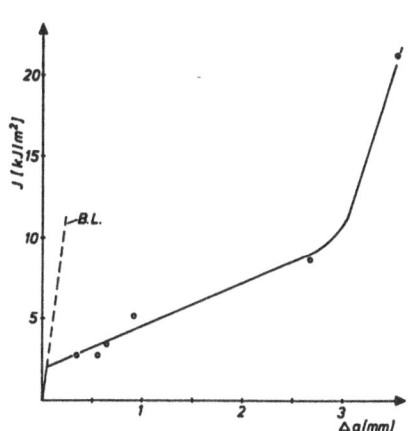

Fig. 5: Crack resistance curve for the determination of a
J_{IC}-value of polycarbonate (J_{IC}=2.06 kJ/m^2) [14]

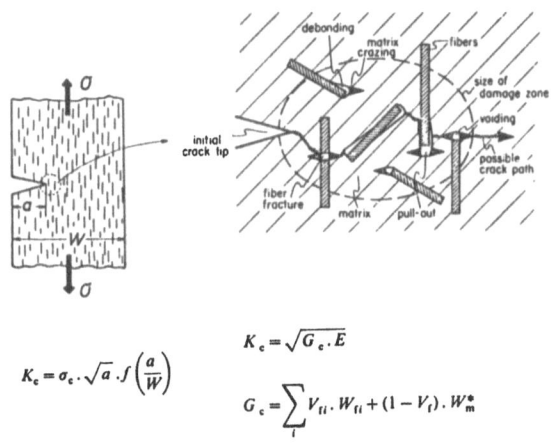

Fig. 6: Schematic illustration of the failure mechanisms during
breakdown of short fiber reinforced thermoplastics, and
corresponding equations for the determination of K_C, the
relationship between fracture toughness and fracture
energy, and the fracture energy, G_C, as the sum of the
partial energy contributions absorbed by the individual
mechanisms

The relative contribution of the individual mechanisms and the absolute value of energy absorbed in each of these processes is highly influenced by different microstructureal factors:

a) fiber length, orientation, volume fraction, type, and distribution over the cross section,

b) deformation behavior of the polymer matrix at certain external testing conditons, and

c) the fiber-matrix bond quality.

Figure 7 reflects an impression which tendencies in fracture toughness as a function of fiber content and crack direction are possible in different polymer/matrix systems. A real comparison between the individual composites can, however, not only be performed on the basis of the fiber volume fraction. For this purpose other, especially fiber related factors must be considered as well. This can be done by the introduction of a reinforcing effectiveness term R, as part of a microstructural efficiency factor, M, which has been introduced and discussed in detail elsewhere [22].

The relative improvement in the toughness of the composite K_{cc} over the fracture toughness of the unreinforced polymer matrix K_{cm}, as plotted against the reinforcing effectiveness term, R, is clearly demonstrated on Fig. 8. In engineering polymers with high fracture toughness, for example polyetheretherketone (PEEK), a fiber reinforcement has at its best no negative effect on the fracture toughness of the composite. In case of a poor fiber/matrix bond quality as for instance valid for PTFE (with very short fiber lengths in addition) a very clear reduction in the fracture toughness of the composite (subscript c) relative to the neat matrix (subscript m) must be expected. In brittle polymer matrices, for example polyphenylene sulfide (PPS) or polyethylene terephthalate (PET) at $-60°C$, fibers are much more effective in improving the fracture toughness than it is tho case for more ductile polymer matrices (e.g. PET at room temperature).

Fracture of Continuous Fiber Reinforced Polymer Composites

a) Fracture Toughness as a Function of Composites Built-up and its Individual Components

The application of fracture mechanics concepts by the use of compact tension (CT)-specimens taken from continuous fiber reinforced composite systems is only then possible, when a quasi-isotropic material behavior precominates. In these cases macroscopically plane

131

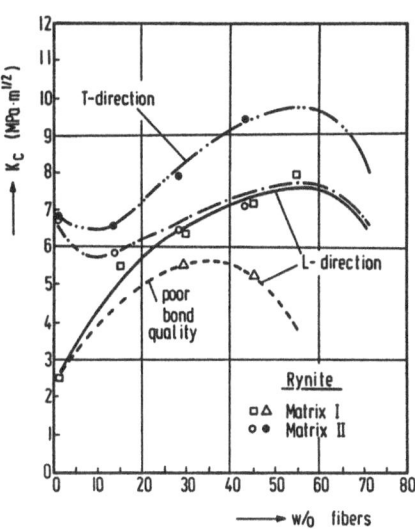

Fig. 7: Fracture toughness as a function of fiber weight fraction
in injection molded, glass fiber reinforced PET plates
(with different fiber matrix bond quality, crack
direction, and initial matrix toughness)

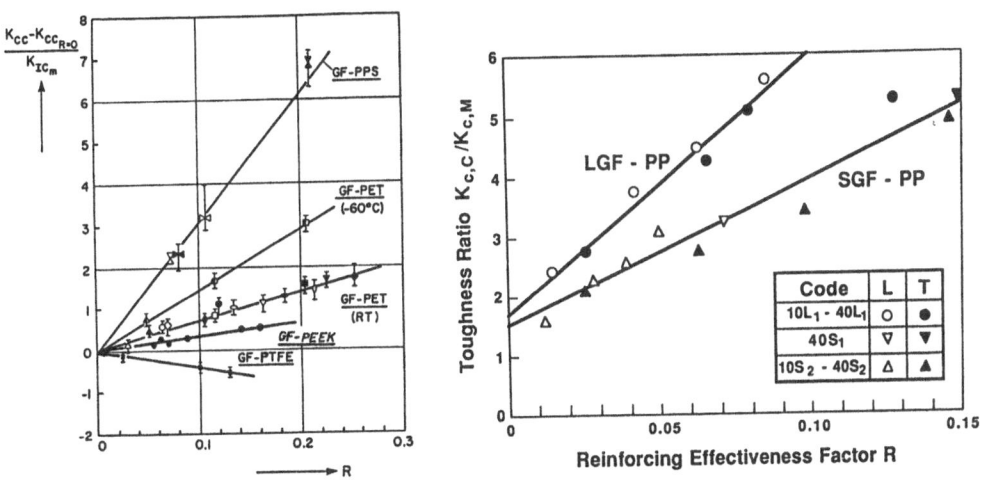

Fig. 8: Influence of the reinforcing effectiveness term, R, on the
variation of the fracture toughness of different
thermoplastic matrices (K_{ICm}) (left) and short and long
fiber reinforced polypropylene (PP)-composites (right)

fractures perpendicular to the applied load axis can be achieved so
that formally fracture toughness values for these materials can be
calculated from the critical load, the initial crack length and the
geometrical correction factor. Table 3 informs about the influence of
various microstructural parameters on the fracture toughness of
composite materials with a quasi-isotropic laminate structure [23].
High strength carbon (C)-fibers result in a higher fracture toughness
than high modulus -C-fibers at a given fiber volume fraction in an
epoxy resin (EP)-matrix. A similar effect on the fracture toughness of
the composite can be expected from an increase in the fiber volume
fraction and/or a quality reduction of the fiber /matrix adhesion. On
the basis of the same laminate built-up, glass fibers (GF) are not as
effective as carbon fibers in enhancing the fracture toughness of an
epoxy matrix composite. For the same type of fiber reinforcement, also
the type of the matrix material can play a dominant role in the
fracture toughness of the composite (compare epoxy and polyimide
(PI)).

b) Interlaminar Fracture Energy

A very special area of weakness in continuous fiber composites opens
up, when cracks start to propagate parallel to the fibers or in the
contact regions between individual lamina (i.e. intra- or interlaminar
fracture). A prevention of this weakness is attempted by the
development of improved fiber/matrix interfaces and new polymer matrix
systems. In the latter case, the replacement of conventional,
relatively brittle epoxy matrices in high performance composites by
new, high temperature resistant, much tougher thermoplastic matrices
seems to be quite promising [24]. Results of the interlaminar fracture
energy obtained for CF/EP- relative to CF/PEEK-composites are plotted
in Figure 9 as a function of the crack tip-opening displacement rate
[25]. It is well established that the much higher values of CF/PEEK
can be related mainly to the larger zone of plastic deformation of the
PEEK-matrix around neighboring fibers in front of the propagation
crack (Fig. 10) [26]. The same conclusions can be drawn from fracture
surface micrographs of the two different composite systems [27].

The actuality of this interlaminar problem is documented in many
publications of recent years, for example [28-30]. Besides the effects
of other microstructural parameters many of these papers also discuss
interlaminar fracture energies obtained under other crack-opening
modes (for example mode II) and under different testing temperatures.

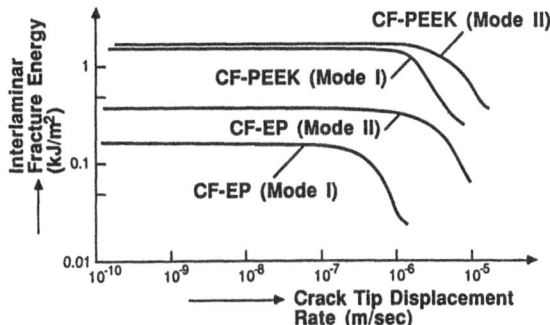

Fig. 9: Interlaminar fracture enrgy G_{Ic} of unidirectional
CF-EP(AS4) and CF-PEEK (APC) laminates as a function of
the crack tip opening rate [25]

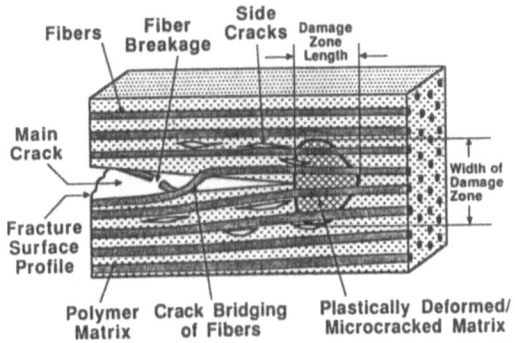

Fig. 10:Size of the damaged area in front of an interlaminar crack
in fiber composites with a brittle resin (left) and with a
ductile polymer matrix (right) [26]

Materials		Fibre volume fraction	Critical Stress Intensity K_c |MPa·m$^{1/2}$|
Type II Carbon/Epoxy	surface treated carbon fibres	0.64	43.3
Type I Carbon/Epoxy		0.64	23.3
Type I Carbon/Epoxy		0.4	18.8
Boron/Epoxy		0.66	48.1
Type I Carbon/Epoxy (untreated fibre)		0.4	48.0
Type II Carbon/Polyimide		0.6	47.8
E-glass/Epoxy		0.6	15.0

Table 3: Fracture toughness of various, quasi-isotropic polymer
composite laminates [23]

CONCLUDING REMARKS

Similar to the conditions known for metallic and ceramic materials, also the fracture behavior and the resulting properties of engineering polymers and their fiber composites are strongly influenced by their microstructures. The individual mechanisms which contribute to the energy absorption during break-down of these materials are, of course, clearly different from those of other materials due to the very complex nature of the polymer morphology and the microstructure of their composites. Nevertheless it is possible at least with polymer composites to achieve values of fracture toughness which fall into the range of typical values measured for metals. This can be considered as a very special advantage, especially when taking into account also the very low density (ρ) of these materials, resulting in a very high value of the specific fracture toughness (K_{IC}/ρ).

ACKNOWLEDGEMENT

The author is grateful to the Deutsche Forschungsgemeinschaft for the financial support of our fracture studies performed with polymer composites (DFG FR 675-2-2).

REFERENCES

[1] K. Friedrich, Progr. Colloid & Polym. Sci. 66 299 (1979).
[2] K. Friedrich, Fortschr. Ber. VDI-Z., Reihe 18, Nr. 18, (1982).
[3] K. Friedrich, Comp. Sci. Technol. 22, 43 (1985).
[4] M.M. Schwartz, Composite Materials Handbook, (Mc. Graw Hill Book Comp., New York, 1983).
[5] K. Schulte, S. Kutter, K. Friedrich, DFVLR-Report No. DFVLR-FB-86-18, 1986.
[6] S. Hashemi, J.G. Williams, J. Mater. Sci.19, 3746 (1984)
[7] W. Grellmann, J.P. Sommer, Akademie der Wiss. der DDR, FMC Series Nr. 17 (1985) 48.
[8] N. Haddaoui, A. Chudnovsky, A. Moet, Polymer 27, 1377 (1986)
[9] K. Friedrich, U.A. Karsch, Fiber Sci. Technol. 18, 37 (1983).
[10] K. Friedrich, Progr. Colloid & Polym. Sci. 64, 103 (1978).
[11] K. Friedrich, in H.H. Kausch (ed.): Crazing in Polymers Springer Verlag, Berlin, 1983) p. 225.
[12] E. Hornbogen, K. Friedrich, J. Mater. Sci. 15, 2175 (1980).
[13] K. Schäfer, K. Friedrich, BEDO-Band Nr. 12, 125 (1979).
[14] J. Krey, R. Walter, K. Friedrich, Z. Werkstofftech. 17, 226 (1986)
[15] J.Karger-Kocsis, K. Friedrich, Polymer 27, 1753 (1986)
[16] J. Karger-Kocsis, K. Friedrich, J. Mater. Sci. 22, 947(1987)
[17] H. Voss, K. Friedrich, J. Mater. Sci. 21, 2889 (1986).
[18] H.H. Kausch, Polymer Fracture (Springer Verlag, Berlin, 1978).

[19] A.J. Kinloch, R.J. Young, Fracture Behaviour of Polymers Appl. Sci. Publ., New York, (1983).

[20] J. G. Wiliams, Fracture Mechanics of Polymers Ellis Harwood Ltd., Chichester, 1984).

[21] K. Friedrich (ed.): Application of Fracture Mechanics to Composite Materials, Elsevier Sci. Publ., Amsterdam, 1989.

[22] H. Voss, Ph.D. Thesis, Technical University Hamburg-Harburg, 1986.

[23] K. Friedrich, E. Hornborgen, A. Sandt, Fortschr. Ber. VDI-Z., Reihe 5, Nr. 82 (1984).

[24] F.N. Cogswell, D.C. Leach, Plastics & Rubber Proc. Appl. 4, 271 (1984)

[25] A.J. Smiley, M.SC. Thesis, Center for Composite Materials, University of Delaware, Report No. CCM-86-01, 1986.

[26] A.J. Klein, Adv. Mater. & Processes 10, 43 (1985).

[27] K. Friedrich, Center for Composite Materials, University of Delaware, Reporet No. CCM 87-52, 1987.

[28] J.W. Gillespie, L.A. Carlsson, A.J. Smiley, Comp. Sci. Technol. 28, 1(1987).

[29] W.. Bradley, W.M. Jordan, Proc. Int. Conf. on Comp. Mater. and Structures, China, June 1986, p. 445.

[30] P. Davies, M.L. Benzeggagh, F.X. De Charentenay, ibid, p. 458.

[31] G.S. Giare, Eng. Fracture Mech. 20, 11 (1984).

MODELLING OF TOUGHENING AND ITS TEMPERATURE
DEPENDENCY IN WHISKER–REINFORCED CERAMICS

Antonios E. Giannakopoulos[*] and Kristin Breder[**]

[*]Department of Solid Mechanics

[**]Department of Physical Metallurgy and Ceramics

The Royal Institute of Technology, S–10044 Stockholm, Sweden

I. Introduction

In recent years, there has been an increasing interest in the mechanical behavior of ceramics. Monolithic ceramics, like Al_2O_3, may show some toughening through microcracking, but the answer to their toughening enhancement is by reinforcing them with rod type of particles.

Reinforced ceramics are designed to be used at severe (aggressive) conditions of high temperatures and corrosive environment. Since they may also undergo extensive microcracking, there is a need for a stable reinforcing material like SiC whiskers. Interfacial properties are very important in connection with the toughening mechanisms present in such composites. The present work examines the constituent and interfacial properties, and models the resulting toughening mechanisms in mode–I, plane strain. Consideration of the thermal residual stresses permits us to include the temperature dependence in the pertinent fracture analysis. This fracture analysis could be useful for temperatures up to the matrix critical temperature.

II. Thermomechanical Properties of Whiskers and Matrix

The whiskers (SiC) are in essence minute, needle–shaped crystals (cubic) of high purity. Statistical analysis showed an average diameter $2a \approx 0.5 \mu m$, and an average length $2L \approx 10 \mu m$. Their mechanical behavior is linear elastic (assumed isotropic), up to their strength limit. Microtensile experiments [15] revealed the exceptional high stiffness (Young's modulus $E_w = 580$ GPa, Poisson's ratio $\nu_w = 0.25$) and high strength $\sigma_{w,uts} = 9$ GPa. Their density is $\rho_w = 3.21$ g/cm^3, and their thermal expansion coefficient is $\alpha_w = 4.8 \times 10^{-6} / ^\circ C$. The SiC whiskers are chemically very stable up to temperatures of $2000^\circ C$.

The matrix (Al_2O_3) is a non–transforming ceramic. This is a brittle polycrystalline aggregate with relatively weak grain boundaries. Regarding its mechanical behavior, intergranular microcracks develop preferably in the direction of the maximum applied principal tensile stress [10]. These microcracks nucleate at grain junctions (i.e.

triple points) in conjunction with extrinsic inhomogeneities (i.e. voids or inclusions). Localized residual stresses form due to thermal expansion anisotropy of the individual grains. After nucleation, microcracks expand along grain facets, motivated by tensile residual stresses, and finally are arrested at neighboring junctions. Microcracking is stable because of the tendency of the residual stresses to alternate between tension and compression on adjacent facet pairs. The formation of such stress–induced microcracks reduces the elastic modulus of the material, resulting in a non–linear stress–strain curve (Fig. 1). Since the material is non–transforming, no permanent strain is expected upon unloading (absence of tranformation dilatation). A load cycle hysteresis develops due to microcrack development only. Detailed knowledge of the microcrack nucleation mechanism is not essential for selecting a nucleation criterion. Briefly, when the grain facet size l' in the material is less than a critical l'_c, stress induced microcracking initiates at a threshold stress σ_0. From thermal analysis [10]

$$\sigma_0 \simeq [(l'_c/l')^{1/2} - 1] G_m \, \Delta\alpha_m \, \Delta T \tag{1}$$

where G_m is the intrinsic shear modulus of the matrix, $\Delta\alpha_m$ is the difference of the thermal coefficients in the grain, and ΔT is the cooling temperature. As the applied stress σ exceeds σ_0, additional grain facets satisfy the microcracking criterion (1) and the microcrack density d increases approximately proportional to $(\sigma-\sigma_0)$. Figure 1 shows that the stress–strain response is linear below σ_0, but thereafter becomes nonlinear since microcracking increases continuously with stress (typical unmicrocracked values for Al_2O_3: $E_m=380$ GPa, $\nu_m=0.25$). The nonlinear loading response resembles that of a plastic material, but since the elastic modulus is reduced by irreversible damage, unloading occurs with a reduced secant modulus. It is further assumed [8] that microcracking saturates, such that the stress–strain response exhibits linearity at large strains. The slope of the microcrack saturated material is

$$E_s/E_m \simeq 1 - (16/9)d_s \tag{2}$$

where d_s is the saturated microcrack density. The saturation part allows the near–tip crack field to be characterized by a stress intensity factor K_t, as we will see in the following.

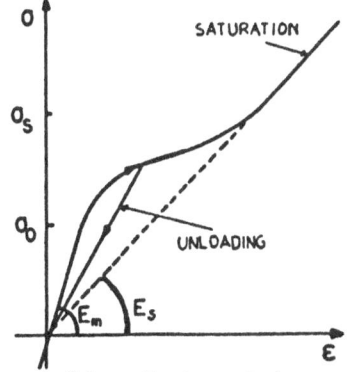

Figure 1: Schematic stress–strain curve for a microcracking ceramic.

The anisotropic microcrack behavior has been extended for the multi–axial stress state [11]. Rate and deformation type of damage constitutive equations can be suitably selected for our purposes. In case of proportional stressing [11]

$$\epsilon_{ij} = \{C^o_{ijkl} + \lambda(\sigma_1) \, n_i n_j n_k n_l\} \, \sigma_{kl} \tag{3}$$

where ϵ_{ij} is the small strain tensor, σ_{kl} is the stress tensor, C^o_{ijkl} is the elastic compliance of the unmicrocracked material, σ_1 is the maximum principal tensile stress, n_i is the direction of σ_1, and $\lambda(\sigma_1)$ is a measure of the cumulative microcracking damage which depends on the uniaxial stress–strain law. The density (Al_2O_3) is $\rho_m = 3.97$ g/cm^3, the thermal expansion coefficient is $\alpha_m = 8.9 \times 10^{-6}/^oC$, and the critical temperature is $T_c \approx 1000^oC$ (rate independent behaviour). Our analysis will be confined to temperatures below T_c.

III. Fabrication, Anisotropy and Residual Stresses

The fabrication of hot–pressed ceramics is described in detail in Ref. [19]. The temperature is increased up to $T_{fabr} \approx 1850^oC$ ($> T_c$), and the pressure (41MPa) is applied along one direction. As a result of the fabrication process the whiskers are randomly distributed, laying on planes normal to the hot–pressed direction, with little clustering.

The 2–D whisker randomness creates anisotropy in the overall elastic moduli (transverse orthotropy), as well as anisotropy in the fracture toughness [3,4]. In the present, we will examine the composite fracture toughness for crack planes parallel to the hot–pressing direction, confining the deformations on planes where the whiskers are randomly distributed.

The effective elastic modulus E can be estimated using the composite cylinder model [6,7]. For $E_w/E_m > 1$ and $\nu_m = \nu_w = \nu = 0.25$, lengthy computations produced Fig. 2. In that figure we summarize the normalized elastic composite modulus E/E_m as a function of the whisker volume percentage c, parametrically with E_w/E_m.

Using the composite cylinder model, we estimate the thermal residual stress acting on the whiskers from the cooling process. The residual stress develops because of a thermal strain mismatch Ω. Within good approximation

$$\Omega = (\alpha_m - \alpha_w)(T_{amb} - T_c) \tag{4}$$

Then the whisker/matrix interfacial pressure p^I is

$$p^I = \frac{E_m}{2\beta} \frac{1-c}{1-\nu} \Omega \ ; \ \beta = 1 - \frac{1-2\nu}{2(1-\nu)} (1 - \frac{cE_w + (1-c)E_m}{E_w}) \tag{5}$$

and under the previous considerations it is always compressive. For temperatures above T_c, the matrix deforms without imposing any residual stresses on the whiskers. Finally it should be noted that the composite density $\rho = c\rho_w + (1-c)\rho_m$ is a decreasing function of the whisker composition and this fact makes the composite attractively lighter than the matrix.

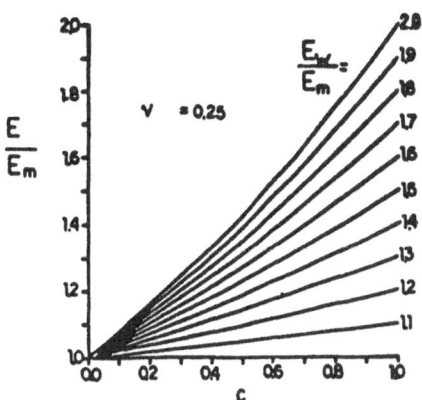

Figure 2: Normalized elastic modulus of the composite.

IV. Interfacial Properties

In the composite system under consideration, there is a very weak adhesive bonding at the matrix/whisker interface (no wetting). Instead, interfacial friction stress is thought to be the important micromechanical property that controls the overall behavior and for this reason it is worth measuring the associated friction coefficient μ_c, assuming a static Coulomb type of friction. A method for measuring the frictional stress is described in Ref [13].

A sharp nano–indenter is applied along the centerline of a long whisker with its axis z normal to a polished surface of the composite halfspace. A force F causes sliding of the whisker along the interface over some distance l. The whisker end slides by a distance u from the surface. The frictional stress τ is related to the applied force and the slip, as well as the thermal residual stress and the friction coefficient, in a non–linear way. In the experiment, the forces F and the sliding displacement u were recorded continuously during loading, unloading, and subsequent loading cycles. The tests can be simulated with FEM [12], parametrized by the friction coefficient μ_c. The computed load–displacement curves were compared with the experimental ones and the best comparison gave the estimate of the friction coefficient ($\mu_c \approx 0.01$).

V. Toughening Mechanisms

A SEM micrograph of the fractured surface is shown in Fig. 3. From that figure and others we may conclude that the possible acting toughening mechanisms are: matrix microcracking (shielding), crack–tip deflection, and whisker pullout (see also [17,18]). A schematic of the acting mechanisms around the crack–tip in mode–I, plane strain conditions is shown in Fig. 4a.

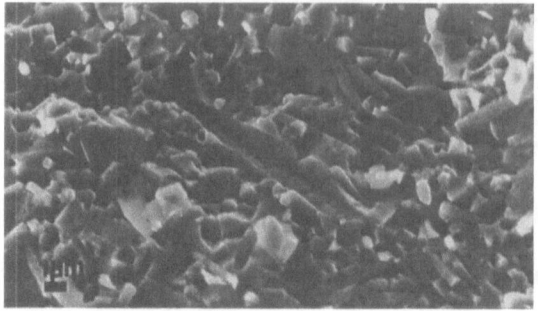

Figure 3: Fracture surface of the composite.

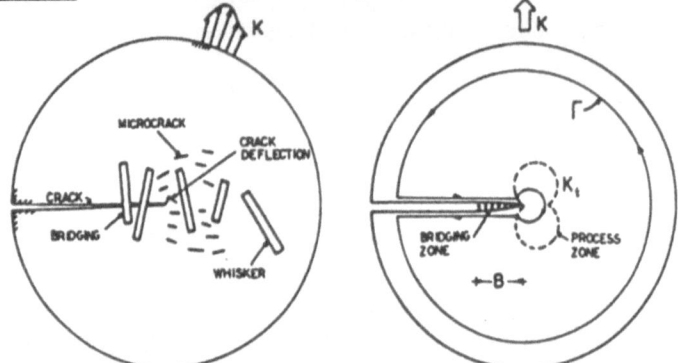

Figure 4: a) Schematic of the toughening mechanisms. b) The path Γ for the calculation of the J–integral.

The matrix microcracking is included directly in the matrix constitutive relation (Eq. (3)). The asymptotic mode–I stress field is [11]

$$\epsilon_{ij} = [C^0_{ijkl} + (1/E_s - 1/E_m)n_i n_j n_k n_l] \; \sigma_{kl} \tag{6}$$

and the asymptotic mode–I stress field is similar to the one given by the linear elastic asymptotic solution [11]. The computed microcrack pattern is shown in Fig. 5.

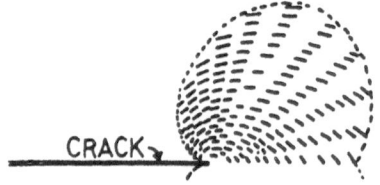

CRACK

Figure 5: Computed microcrack pattern.

Crack deflection accounts for the toughening increase due to a local increase of the crack surface. The toughening ratio D of the applied energy release rate over the energy release rate due to crack deflection is shown in Fig. 6, as a function of the whisker volume concentration c and aspect ratio L/a. The results were computed using 2–D averaging of the local twist and tilt of the crack front as in Ref. [9].

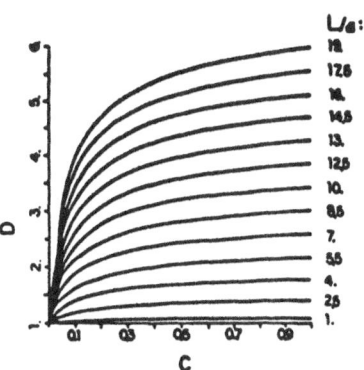

Figure 6: Toughening due to deflection.

We contemplate the following fracture sequence. The matrix fractures at the crack–tip whereas simultaneously whiskers are pulled out of the matrix behind the tip, impeding the crack opening and therefore enchancing the composite toughness. We model this behavior by introducing at the tip region a so called bridging zone [5] (Fig. 7a). The zone is assumed to consist of "springs" with a prescribed relation between their stress σ_b and the crack opening displacement v, and stems from a micromechanical analysis. The development of the bridging zone is connected with the R–curve and will not be addressed here. We will confine our attention to the steady state, small scale bridging. That is to say, the bridge length B is constant and moves together with the crack–tip in a self–similar way. At the tip, because of the severe local tensile stress field and remembering that the frictional interface calls for compression, the whiskers remain in the average intact, emanating as the tip advances. Simultaneously the whiskers at the tail of the bridging zone are completely pulled out (the bridging zone is considered to be completely included in the near–tip field). At the vicinity of the crack faces, only the clamping residual stress p^I is essentially acting on the pulled whiskers (Fig. 7b). Then, their average σ_b–v behavior is approximated by a frictional type of pullout law, neglecting the elastic hysteresis (due to high stiffness), as shown in Fig. 7c.

$$\sigma = \frac{L}{2\,\pi\,a}\,c\,\tau\left(1 - \frac{2}{L}\,v\right) \qquad (7)$$

$$\tau = \mu_c|p^I| \qquad \sigma^* = \frac{L}{2\,\pi\,a}\,c\,\mu_c|p^I| \qquad v^* = L/2$$

$$V = \sigma^*\,v^* \qquad (8)$$

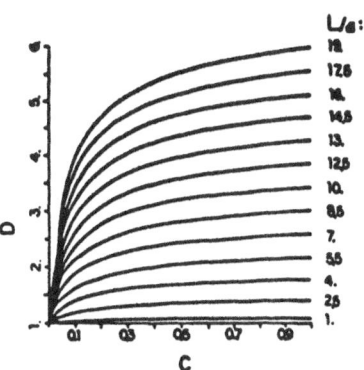

Figure 7: The pullout mechanism.

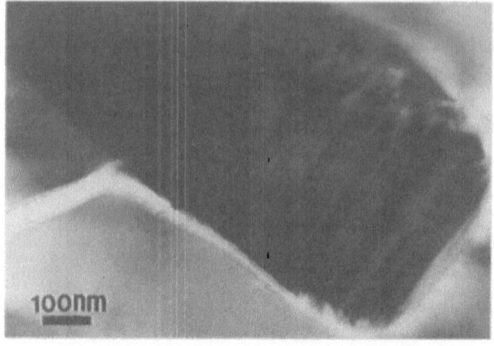

Figure 8: Transmission electron micrograph of pulled-out whiskers.

VI. Synergism

Denote the asymptotic remote stress field around the crack–tip (Fig. 4) by

$$\sigma_{\alpha\beta} = \frac{K}{(2\pi r)^{1/2}} f_{\alpha\beta}(\theta) \tag{9}$$

where K is the applied mode–I stress intensity factor that includes the geometry and loading effects in the usual way it is assumed in small scale analysis. The near–tip stress field will be of the same form as in Eq.(9), but with K_t in place of K . For monotonic loading, FEM results showed that the stress paths are nearly proportional. In addition, the region of validity of the near–tip asymptotic was also shown to be substantial, so autonomy with respect to K_t exists. This is essential to assertain the proposed stress intensity factor fracture criterion [14].

Under the previous considerations we can apply the path independent J–integral [16] for a path Γ as shown in Fig. 4b,

$$J = \int_\Gamma (W\,\eta_1 - \sigma_{\alpha\beta}\,u_{\alpha,1}\,\eta_\beta)\,ds \tag{10}$$

in order to calculate the relation between K and K_t, or equivalently the toughness K/K_t (K_t is an intrinsic toughness). In Eq. (10), W is the strain energy density (nonlinear due to damage), u_α is the displacement vector, and η_α is the unit vector normal to the line Γ. Eq. (10) yields the following energy release rate balance

$$\frac{1-\nu^2}{E} K^2 = (1-c)\,\frac{1-\nu_m^2}{E_m}\,[1 + \frac{15\pi + 56}{30\,\pi}(E_m/E_s - 1)]\, D\, K_t^2 + V \tag{11}$$

The left hand side of Eq. (11) is the energy release rate due to the far field. The right side of (11) has two contributions (related with the rule of mixture). The first comes from the matrix that is microcracking (the extra term in the bracket), and deflecting (D coefficient). The second contribution (V) comes from the energy dissipated due to pullout (pseudo–ductility). The cross effect between whiskers and microcracks was neglected, implying a dilute approximation.

The set–up of Eq. (11) is very flexible in the sense that one may simply discard any individual toughening mechanism where evidence is little or incomplete. Furthermore, the temperature effect is directly incorporated through V (Eq. (8)). The temperature dependence of the elastic moduli is very weak for the temperature ranges under consideration.

VII. Some Approximations

It is instructive at this point to reduce the complexity of the previous formulation. We now consider complete linear elastic behavior, but with the bridging mechanism acting. Then, from the equation of the asymptotically (far field) parabolic crack profile we may estimate the bridging length B from the displacement condition at the tail of the bridging zone

$$v^* = 4 \, K \, (B/2\pi)^{1/2} \frac{1 - \nu^2}{E} \tag{12}$$

For the Al_2O_3/SiC we find $B \simeq 3 - 6$ μm, which is consistent with observations. Note that however small, the bridging zone may contribute to the toughening K/K_t

$$\frac{K}{K_t} = (1 - 2C_1)^{-1/2} \quad ; C_1 = \frac{\sigma^* \, (2B)^{1/2}}{K \, \pi^{1/2}} \, , \quad 0 < C_1 < 1/2 \tag{13}$$

The normalized bridging stress distribution $\sigma(r/B)/\sigma^*$, acting on the crack faces behind the crack–tip, can be computed from

$$1 - \sigma/\sigma^* = (r/B)^{1/2} - C_1 \int_0^1 (\sigma/\sigma^*) \log \frac{(r/B)^{1/2} + (r')^{1/2}}{|r/B - r'|^{1/2}} \, dr' \tag{14}$$

where $0<r/B<1$ is the normalized position along the bridging zone starting from the crack–tip. Note that Eq. (14) is a disguised equation of the crack opening profile: the initial parabolic profile is closing due to the effect of the whisker pullout opposing the crack opening. Equation (14) is a Fredholm equation of the second kind. It was solved with FEM by taking its weak form, discretizing σ/σ^* in r/B, and performing the necessary integrations. The results for σ/σ^* as function of r/B are shown in Fig. 9, parametrized with C_1. Note that Fig. 9 can be used to present the near–tip crack opening profile as well (from Eq. (7)).

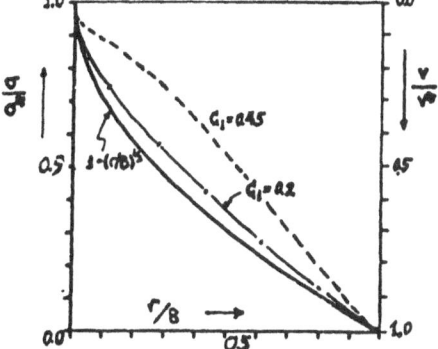

Figure 9: The normalized bridging stress distribution.

VIII. Examples and Conclusions

As an example, we examined an Al_2O_3/SiC system. Without going into further details, we were able to insert actual independent data for the variables of our model and compute the toughness as a function of the whisker composition c. The computed and the experimental results [1] are shown in Fig. 10. For the same material data, the ambient temperature dependence of toughness was calculated for temperatures up to the matrix critical temperature. The computed and the experimental results [2] are shown in Fig. 11.

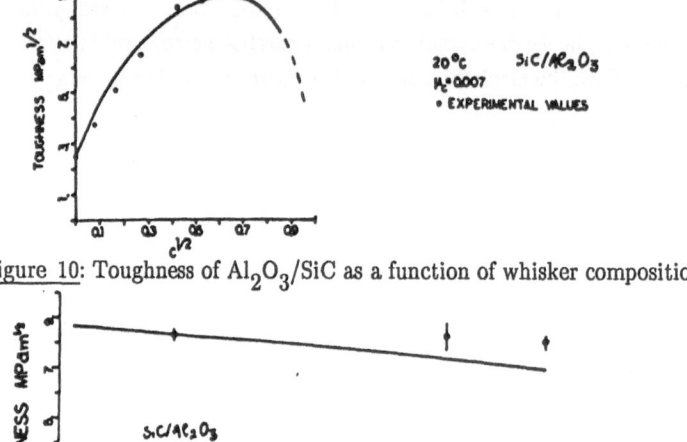

Figure 10: Toughness of Al_2O_3/SiC as a function of whisker composition c.

Figure 11: Toughness of Al_2O_3 as a function of ambient temperature.

All the predictions agree very well with the available experimental data. From the presented examples we can see how knowledge of the key micromechanical parameters (i.e. c, μ_c, d_s, etc) and the constituent properties affect the composite toughness. It is worth mentioning the optimality in toughness with respect to the whisker volume concentration. Note also the mild decrease of toughness with temperature, indicating the small effect of the pullout mechanism, compared with the other toughening mechanisms. However, pullout mechanism may be enhanced, if we could dictate an increased friction and/or aspect ratio.

Aknowledgements

The authors would like to thank Mr K. Zeng for the photographs shown in Figs. 3, 8.

References

1) P. F. Becher, C–H. Hsueh, P. Angelini and T. N. Tiegs, "Toughening Behavior in Whisker–Reinforced Ceramic Matrix Composites", J. Am. Ceram. Soc., **71** [12], 1050–1061 (1988).

2) P. F. Becher and T. N. Tiegs, "Temperature Dependence of Strengthening by Whisker Reinforcement: SiC Whisker Reinforced Alumina in Air", Adv. Ceram. Mater., **3** [2], 148–154 (1988).

3) K. Breder, D. Rowcliffe, K. Zeng and A. E. Giannakopoulos, "Nano–indentation Study of Interfaces in Ceramics", 91st An. Meeting of the Am. Ceram. Soc., Indian. (1989).

4) K. Breder, K. Zeng and D. Rowcliffe, "Indentation Testing of an Al_2O_3/SiC wisker Composite", Ceram. Eng. Sci. Proc., **10** (1989).

5) B. Budiansky, "Micromechanics II", pp. 25–32, in Proc. Tenth U.S. National Congr. Appl. Mech., Austin, Texas (1986).

6) R. M. Christensen and K. H. Lo, "Solutions for Effective Shear Properties in Three Phase Sphere and Cylinder Models", J. Mech. Phys. Solids, **27**, 315–330 (1979).

7) R. M. Christensen and F. M. Waals, "Effective Stiffness of Randomly Oriented Fibre Composites", J. Composite Materials, **6**, 518–532 (1972).

8) A. G. Evans and Y. Fu, "Some Effects of Microcracks on the Mechanical Properties of Brittle Solids –II. Microcrack Toughening", Acta metall., **33** [8], 1525–1531 (1985).

9) K. T. Faber and A. G. Evans, "Crack–Deflection Processes –I. Theory", Acta metall., **31** [4], 565–576 (1983).

10) Y. Fu and A. G. Evans, "Some Effects of Microcracks on the Mechanical Properties of Brittle Solids –I. Stress, Strain Relations", Acta metall., **33** [8], 1515–1523 (1985).

11) A. E. Giannakopoulos, "Fracture Mechanics of Monolithic Ceramic Materials under Small Scale Microcracking Conditions", Ph.D. thesis, Brown University (1989).

12) A. E. Giannakopoulos, "Finite Element Simulation of Fiber Indentation Experiments in Composites with Coulomb Friction Interface", report no.116 (1989), Dept of Solid Mech., The Royal Institute of Technology, Stockholm, Sweden.

13) D. B. Marshall, "Interfaces in Ceramic Fiber Composites", in Ceramic Microstructures '86: Role of Interfaces, 859–868, eds J. A. Pask and A. G. Evans, Plenum, N.Y.

14) F. Nilsson, private communication.

15) J. J. Petrovic, J. V. Milewski, D. L. Rohr and F. D. Gac, "Tensile Mechanical Properties of SiC Whiskers", J. Mat. Science, **20**, 1167–1177 (1985).

16) J. R. Rice, "Mathematical Analysis in the Mechanics of Fracture", in Fracture, ed. H. Liebowitz, vol. 2, 191–311 (1968).

17) R.Rice, "Ceramic Matrix Composite Toughening Mechanisms: An Update", Ceram. Eng. Sci. Poc., **6**, 589–607 (1985).

18) M. Rühle, D. B. Dalgleish and A. G. Evans, "On the Toughening of Ceramics by Whiskers", Scripta Metallurgica, **21**, 681–686 (1987).

19) G. C. Wei and P. F. Becher, "Development of SiC–Whisker–Reinforced Ceramics", Am. Ceram. Soc. Bull., **4** [2], 298–304 (1985).

THEORETICAL MODELS FOR THE EXPLANATION OF TOUGHENING MECHANISMS IN HEAVY-DUTY-CERAMICS

Wolfgang H. Müller[†]
Hermann-Föttinger-Institut der TU-Berlin
Straße des 17. Juni 135, D-1000 Berlin 12

1. Introduction

Ceramic materials show several remarkable properties: They can sustain very high temperatures, they do not corrode easily, even when subjected to an aggressive chemical environment, and they are extremely wear-resistant. Thus, they could become an interesting alternative to metals in the near future and replace these in many industrial applications like engine parts, roller bearings or cutting tools. Moreover, most ceramics are biocompatible and, indeed, modern surgery uses them already as bone transplants /1/.

Unfortunately, the fracture resistance of monolithic engineering ceramics (i.e. parts made completely of e.g. Mullite, Cordierite or Al_2O_3) is usually very small, when compared to other materials: K_{IC} (steel) ~ 40-60 MPam$^{1/2}$, K_{IC} (cast iron) ~ 10-15 MPam$^{1/2}$, K_{IC} (Al_2O_3) ~ 2-4 MPam$^{1/2}$. It goes without saying, that such a low fracture toughness is clearly not sufficient for the above mentioned technical purposes.

However, the K_{IC} of ceramic <u>composites</u> may be much higher than the toughness of its individual components. Two composite-systems seem to be especially promising: Ceramics which are toughened by addition of Zirconia (ZrO_2) particles, and ceramics which are reinforced by addition of SiC-whiskers and fibers. The following experimental data may serve as an orientation: K_{IC} (Al_2O_3 + ZrO_2) ~ 6-20 MPam$^{1/2}$, K_{IC} (Mullite + SiC-whisker) ~ 6-9 MPam$^{1/2}$ (/2-4/). In the following chapter we shall briefly explain several micromechanisms which are responsible for the resulting increase in toughness of these composites.

[†] Visiting Scholar, Stanford University, January 1990 - June 1990

2. Toughening Mechanisms in Ceramic Composites

a) Stress Induced Transformation Toughening: If bulk Zirconia is cooled down to room temperature its crystal lattice will transform abruptly at 1200 °C from a tetragonal into a monoclinic configuration. In fact, this phase transition leads macroscopically to an increase in volume of about 3%. Thus, if Zirconia particles (μm-size) are embedded in a ceramic matrix, a constraining pressure will be exerted whenever they start to transform so that, even at room temperature, Zirconia inclusions can be stabilized in their tetragonal variant.

If, however, a crack in the neighborhood of the particles is made to extend under load, the matrix constraint will be released due to the enormous stress-concentrations around the crack tip, and the particles switch into monoclinic symmetry. Because of the volume expansion a compressive process zone will now be generated around the crack. Thus, *extra work would be required to move the crack through the ceramic accounting for the increase in toughness and hence strength /3/.*

b) Crack shielding: If a crack in an Mullite or Cordierite matrix runs towards a SiC-whisker or fiber, its tip will be "shielded and blunted", since the SiC-component is usually much stiffer than its surroundings: $E(Cordierite)/E(SiC) \sim 1/3$. This in turn will macroscopically lead to an increase in K_{IC}.

c) Debonding: Due to the enormous stress concentrations at crack tips it may happen that a crack separates on its way through a ceramic whiskers or fibers from the surrounding matrix. And this, of course, goes to the expense of its own energy.

d) Pullout: For similar reasons as in c) crack propagation may result in the pullout of fibers and whiskers, which again reduces the energy of the cracks and will finally lead to an increase in toughness.

We will now present several simple models which allow a quantitative understanding of the above mentioned energy-dissipative micromechanisms a)-c), within the framework of linear elasticity. For a discussion of the fiber-pullout the reader is referred to the article "Static and Dynamic Pullout of an Elastic Rod from a Rigid Wall" by I. Müller et al in this volume.

3. Calculation of the Increase in Toughness ΔK_{IC}.

We start with Griffith's fracture criterion which for the crack shown in Fig.1 reads:

$$K_I|_c = K_{IC} \, (matrix) \tag{3.1}$$

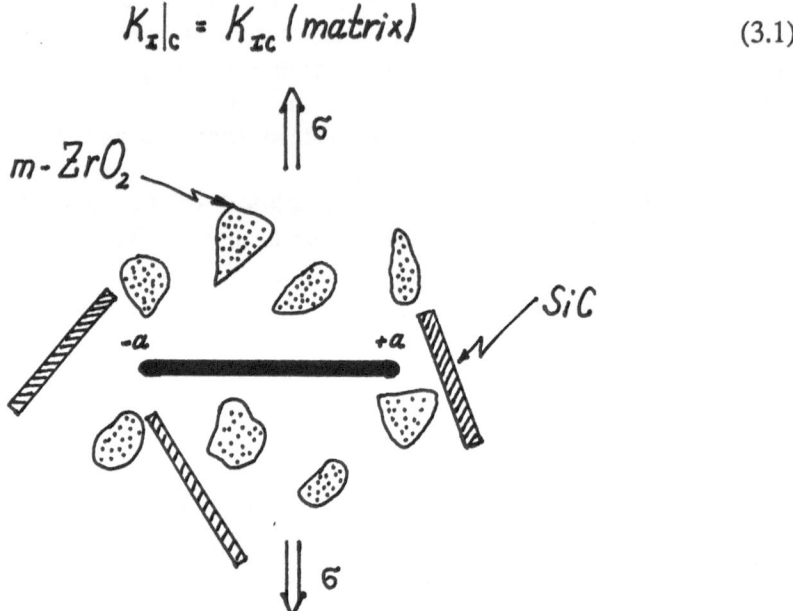

Fig. 1: Griffith crack surrounded by monoclinic Zirconia and SiC-whiskers

If the crack were not surrounded by transformed Zirconia particles or SiC whiskers the K_I would simply be given by:

$$K_I = \sigma \sqrt{\pi a} \tag{3.2}$$

But due to the presence of the particles and whiskers this expression has to be changed into:

$$K_I = \sigma \sqrt{\pi a} \, Y(...) \tag{3.3}$$

where the correction function Y(....) depends in a complicated manner upon the considered geometry , the different elastic constants of the matrix and the inclusions, and upon the volume expansion of the ZrO_2. Or in other words: Y(...) reflects the strength of transformation toughening as well as of crack shielding. In order to make this more explicit we combine (3.1-3.3) and get for the critical external stress σ_c in the monolithic ceramic:

$$\sigma_c = \frac{K_{Ic}(matrix)}{\sqrt{\pi a}} \tag{3.4}$$

and in the composite material:

$$\sigma_c = \frac{K_{Ic}(matrix)}{\sqrt{\pi a}} Y^{-1}(...) \tag{3.5}$$

Thus, if the inverse of the correction function $Y^{-1}(...)$ is greater than 1, a higher stress value must be applied in order to destabilize the system. This is exactly what we expect if the transformed Zirconia particles are above or below the flanks of the crack, so that they compress them, or if the crack is in front of a stiff SiC-whisker. However, if the crack is in front of a transformed ZrO_2 inclusion, its flanks will be opened due to the volume expansion of the Zirconia. Thus, it will be destabilized and run into the particle. In this case $Y^{-1}(...)$ should become smaller than 1 and a smaller load σ_c is necessary for destabilization. Note, that if the crack has run into the transformed particle its flanks will be again under compression. So finally $Y^{-1}(...)$ will always be greater than 1 and we may calculate the increase in toughness according to:

$$K_{Ic}(matrix + \frac{ZrO_2}{SiC}) := \sigma_c \sqrt{\pi a} \equiv K_{Ic}(matrix) Y^{-1}(...) \tag{3.6}$$

4. Methods for the Calculation of the Correction Function Y(...)

In order to evaluate (3.6) it is obviously necessary to know the correction function Y(...). Its determination is generally a non-trivial problem of linear elasticity. In the following we present two methods for its calculation and several corresponding numerical results.

4.1 The Integral Equation Technique

4.1.1 Analysis

We consider the following two situations (see Fig.2): An arbitrarily oriented Griffith crack interacting with a transformed ZrO_2 - inclusion and a pressurized Griffith crack perpendicular to a SiC-whisker.

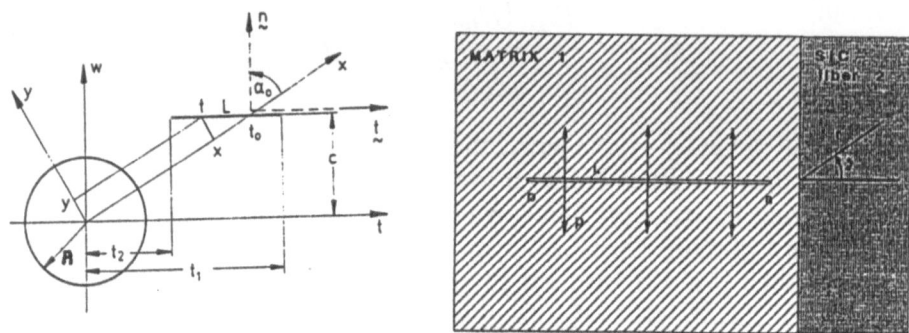

Fig. 2: Crack geometries under consideration

Erdogan and al /5,6/ have shown how to calculate Y(...) in such cases. The idea is to simulate the crack by a continuous array of dislocations, the distribution of which will be determined from the fact that the flanks of the crack must be free of forces. This consideration leads to systems of Cauchy integral equations which, in general, must be solved numerically. E.g. one finds for the case of an arbitrarily oriented matrix crack together with a transformed Zirconia particle:

$$\int_{t_2}^{t_1} g(t_o) \frac{c}{\sqrt{c^2+t_o^2}} \frac{dt_o}{t_o-t} + \int_{t_2}^{t_1} k_{11}(t,t_o) g(t_o) dt_o +$$

$$+ \int_{t_2}^{t_1} f(t_o) \frac{t_o}{\sqrt{c^2+t_o^2}} \frac{dt_o}{t_o-t} + \int_{t_2}^{t_1} k_{12}(t,t_o) f(t_o) dt_o = -\frac{\pi(\kappa_1+1)}{2\mu_1} p_1(t)$$

$$\int_{t_2}^{t_1} g(t_o) \frac{t_o}{\sqrt{c^2+t_o^2}} \frac{dt_o}{t_o-t} + \int_{t_2}^{t_1} k_{21}(t,t_o) g(t_o) dt_o +$$

$$- \int_{t_2}^{t_1} f(t_o) \frac{c}{\sqrt{c^2+t_o^2}} \frac{dt_o}{t_o-t} + \int_{t_2}^{t_1} k_{22}(t,t_o) f(t_o) dt_o = -\frac{\pi(\kappa_1+1)}{2\mu_1} p_2(t)$$

(4.1)

where $g(t_0)$ and $f(t_0)$ denote the components of the Burger's vector for the unknown distribution of dislocations. k_{ij} are extremely complicated integral kernels whose explicit form can be found e.g. in /5,7/. p_i designates the normal and tangential forces of a virtual crack in the undamaged material (see /7/), i.e. p_i takes account of the external stresses as well as of the internal forces which are due to the phase transformation.

Having solved (4.1) the stress intensity or in other words the correction function will be calculated as follows, e.g.:

$$g(t) = w(t) G(t), \quad f(t) = w(t) F(t), \quad w(t) = (t - t_2)^{-\frac{1}{2}} (t_1 - t)^{-\frac{1}{2}}$$

$$K_I(t_1) = - \frac{2\mu_1}{1 + \kappa_1} \frac{\pi^{1/2}}{\sqrt{a(c^2 + t_1^2)'}} \left(c\, G(t_1) + t_1\, F(t_1) \right)$$

(4.2)

4.1.2. Results

Fig. 3. shows the correction functions Y(...) for cracks which interact with transformed Zirconia particles in an Alumina matrix or with stiff SiC-whiskers. The plots are the result of a numerical treatment of the above mentioned integral equations. For details the reader is referred to /7-9/. Transformation toughening and crack shielding are both clearly visible: If the crack is above or in the ZrO_2-particle the correction function Y(...) is smaller than 1. The same holds for a crack in a matrix which is much softer than the embedded SiC-whisker.

It should be noted that by means of the integral equation technique it is not only possible to calculate the stress-intensities but the stresses as well. This has been done in /8/ in order to determine the forces acting perpendicularly and tangentially on a SiC whisker as function of the stiffness of the surrounding matrix. These results can help to understand the inset of the debonding of fibers.

4.2 The Laurent Series Approach

4.2.1 Analysis

The method of integral equations allowed to consider the interaction of one transformed particle with a crack. In order to study the influence of several particles we shall idealize them and treat them as pressurized holes using Isida's Laurent series expansion technique

/10/. Although there is some influence from the difference between the elastic constants of the Zirconia and the surrounding Alumina, this idealization seems to be justified, since the influence is neglegibly

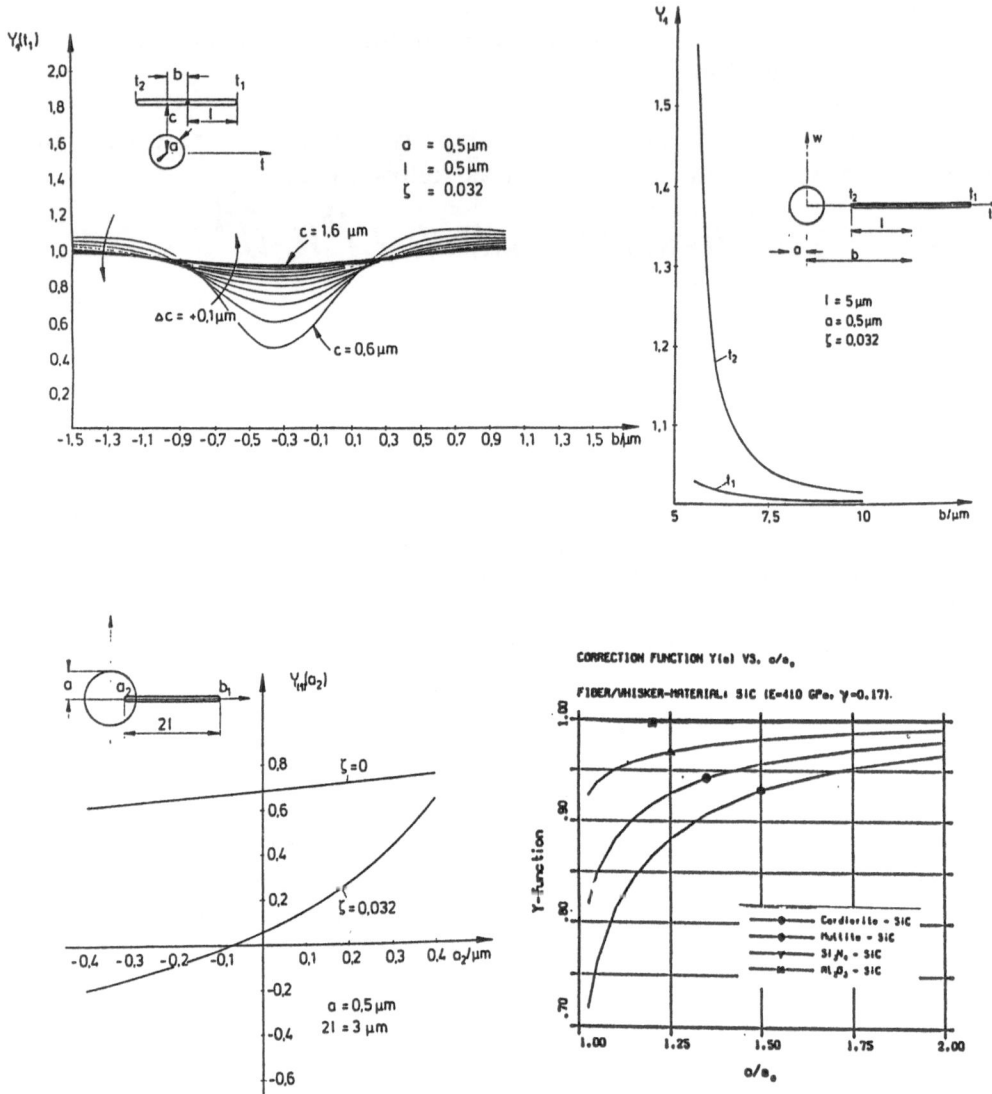

Fig. 3: Correction functions calculated with the integral equation technique taken from /7-9/

small when compared to the phase transition on the correction function /7/.

Isida starts with the Muskhelishvili-Kolosov equations (where d has been introduced as a suitable reference length):

$$t_{xx} + t_{yy} + \frac{8\mu}{1+\kappa}\, \omega_{xy} = 4\sigma\, \phi'(z)$$

$$t_{xx} - t_{yy} - 2i\, t_{xy} = -2\sigma\, \{\bar{z}\, \phi''(z) + \psi''(z)\} \tag{4.3}$$

$$2\mu\,(u_x - i\,u_y) = \sigma d\, \{\kappa\bar{\phi}(\bar{z}) - \bar{z}\phi'(z) - \psi'(z)\}$$

and expands the Goursat functions ϕ and ψ into Laurent series:

$$\phi(z) = \sum_{n=0}^{\infty} \{ F_n\, z^{-(n+1)} + M_n\, z^{n+1} \}$$

$$\psi(z) = -D_0\, \ln z + \sum_{n=1}^{\infty} D_n\, z^{-n} + \sum_{n=0}^{\infty} K_n\, z^{n+2} \tag{4.4}$$

The appearing coefficients are then determined from the boundary conditions (for the case of an extension of Isida's equations to the case of pressurized holes see /11/) and finally the correction function is calculated using Sih's formula:

$$K_{\mathbb{I}} - i\, K_{\mathbb{II}} = 2\sigma\, (2\pi d)^{1/2} \lim_{z \to a/d} \{ (z-a/d)^{1/2}\, \phi'(z) \} \tag{4.5}$$

4.2.2. Results

Fig. 4 shows some typical results for the correction function, which are taken from /11/. One clearly detects that two particles are more effective in stabilizing the crack, provided that they are properly distributed above and under its tips. A statistical evaluation shows that in the case of two particles the resulting local toughness K_{IC} (calculated acc. to (3.6)) can become 7 MPam$^{1/2}$ whereas just as before in the case of the integral equation technique one particle leads to an average K_{IC} in the order of 5 MPam$^{1/2}$.

5. Conclusion and Outlook

The survey presented in the present paper has shown that it is nowadays possible to investigate and predict the mechanical properties of modern engineering ceramics by means of adequate

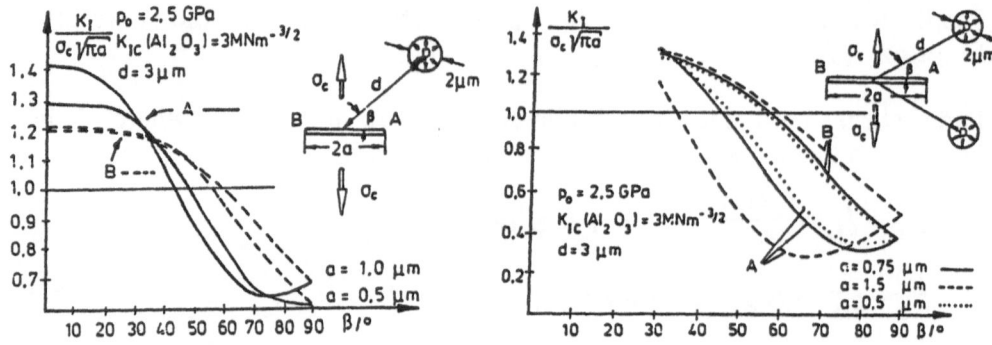

Fig. 4: Correction functions for transformation toughened Alumina calculated by means of the Laurent series expansion technique

numerical techniques. It can be expected that these techniques will be the basis of computer expert systems for ceramic materials in the near future.

6. Acknowledgement

This work was supported by a Max Kade Foundation grant, which is herewith gratefully acknowledged.

7. References

/1/ *Technische Keramik*, Vulkan Verlag, Essen, 1988

/2/ Heuer, A.G., *Transformation Toughening in ZrO_2-Containing Ceramics*, J. Amer. Ceram Soc., **70**, 689, 1987

/3/ Stevens, R., *An Introduction to Zirconia*, Magnesium Elektron Publication No. 113, 2nd edition, 1986

/4/ Claussen, N., Petzow, G., *Whisker Reinforced Oxide Ceramics*, Journal de Physique, **47**, C1-693, 1986

/5/ Erdogan, F., Gupta, G.D., Ratwani, M., *Interaction Between a Circular Inclusion and an Arbitrarily Oriented Crack*, J. Appl. Mech., 1007, 1974

/6/ Cook, T.S., Erdogan, F., *Stresses in Bonded Materials with a crack Perpendicular to the Interface*, Int. J. Engng. Sci., **10**, 677, 1972

/7/ Müller, W.H., *The Exact Calculation of Stress Intensity Factors in Transformation Toughened Ceramics by Means of Integral Equations*, Int. J. Fract., **41**, 1, 1989

/8/ Müller, W.H., *Toughening Mechanisms in Heavy-Duty-Ceramics*, in Proc. ECerS I conference, Maastricht, 1989, in print

/9/ Müller, W.H., *Simulation of Crack Propagation in Particle and Fiber-Reinforced Ceramics*, in Proc. Ceramic-Ceramic-Composites conference, Mons, 1989, in print

/10/ Isida, M., *Method of Laurent Series Expansion for Internal Crack Problems*, Mechanics of Fracture I, Methods of Analysis and Solutions of Crack Problems, ed. G.C. Sih, Noordhoff, Leyden, 1973

/11/ Müller, W.H., *Numerical Methods for the Description of "Ceramic Steels" (ZrO_2-Containing Ceramics)* , in Fracture Control of Engineering Structures, III, 1986

THERMO–MECHANICAL FATIGUE
OF SHAPE MEMORY ALLOYS

Erhard Hornbogen

Institut für Werkstoffe, Ruhr–Universität Bochum

Postfach 10 21 48, 4630 Bochum 1, F. R. Germany

Abstract

Fatigue of shape memory alloys can be caused by thermal cycling through the range of martensitic transformation or by mechanical cycling. The effects differ depending on the temperature range: stable high temperature phase (austenite), or low temperature phase (martensite), or the range where stress induced transformation can take place. A distinction is made between mechanical fatigue and shape–memory fatigue, i.e. hardening and crack formation, or a shift in transformation temperatures and loss of memory. The microstructural origin of all phenomena is discussed.

Introduction

Shape memory alloys show a number of properties which are qualitatively different from those of all other solids. This is also true for the response to thermal and/or mechanical cycling. The microstructural aspects of fracture of normal alloys have recently been treated in a systematic way.[1] The number of papers on shape memory alloys was not large in the past.[2-5] Presently, there is a systematic effort in this field especially on CuZn– and FeMn–base alloys.[6-11]

Shape memory in alloys is connected with a crystallographic transformation of a high temperature phase β (austenite) into martensite α_M. This implies that under certain conditions of stress σ or temperature T both phases can coexist, while a mono–phase structure of β exists at $T \geq A_f$ and of α_M at $T \leq M_f$. Figure 1a presents the thermo–dynamic background for this phase transformation and 1b explains the designations of the various temperatures, which are encountered during a transformation cycle. The martensitic transformation is usually of first order and heterogeneously nucleated (structural defects, surface). In the temperature–range $\Delta T = T_0 - M_s$ second order phenomena may occur, which are known as pre–martensitic (spinodal shear waves).[12]

Figure 2 compares the normal stress σ, strain ϵ, temperature T behavior of materials with the three particular effects of shape memory alloys. The occurrence of the two–way effect requires

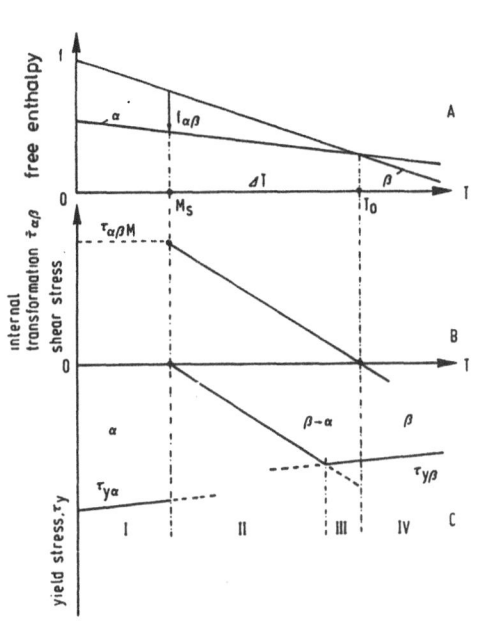

Figure 1a: Discussion of thermo–dynamic and mechanical properties of the high temperature phase β, schematic. a) Temperature dependence of the free energy of the phases β and α (same chemical composition); b) at $T < T_0$ an internal shear stress $\tau_{\beta\alpha}$ originates which leads to transformation at $T = M_s$; c) the yield stress $\tau_{y\beta}$ of β is reduced by $\tau_{\beta\alpha}$. Four temperature ranges: I martensite, II stress–induced martensite α, III plastic strain–induced martensite, IV stable austenite β.

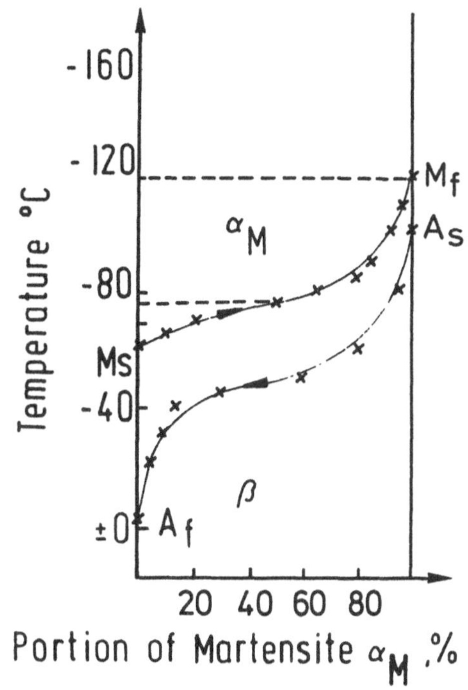

Figure 1b: Course of martensitic transformation and reverse transformation of β–CuZn from X–ray intensities, (200)–β_1–reflections, 60.80 wt.–% Cu). M_s – martensite start; M_f – martensite finish; A_s – austenite start; A_f – austenite finish.

"training" of the material. This implies the introduction of internal stresses or nuclei which favour certain crystallographic shear systems for the $\beta \rightarrow \alpha_M$ transformation.

The limit of the bulk shape change is determined by this crystallographic shear. The memory is due to the fact that the reverse transformation $\alpha_M \rightarrow \beta$ takes the same path as the martensitic one. This phase usually shows a higher degree of microstructural disorder: i. e. a single–crystal or grain of β is transformed into a poly–crystal of α_M. This, in turn, permits the quasi–plastic deformation during the one–way treatment (Fig. 2 and 3).

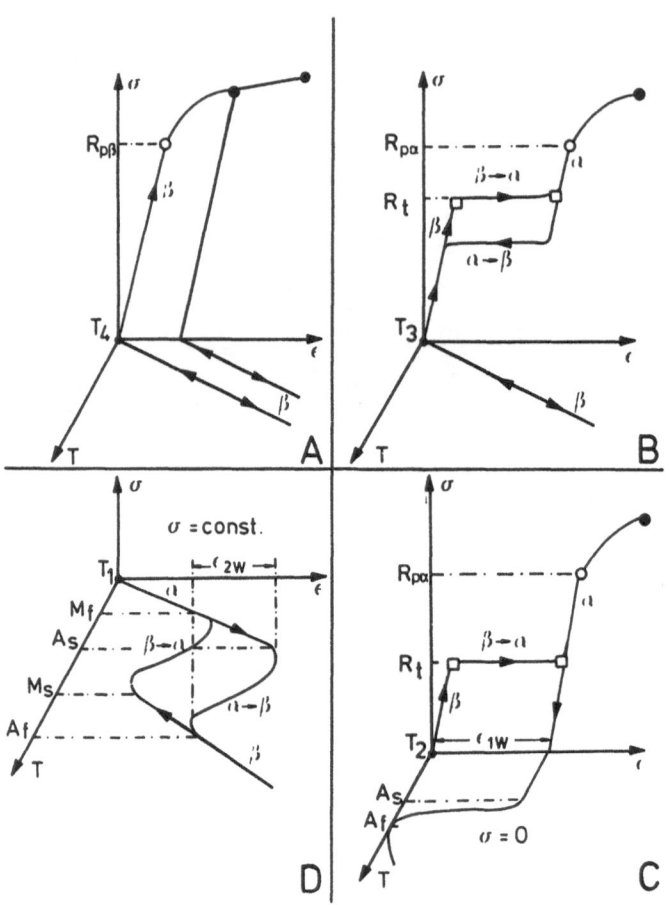

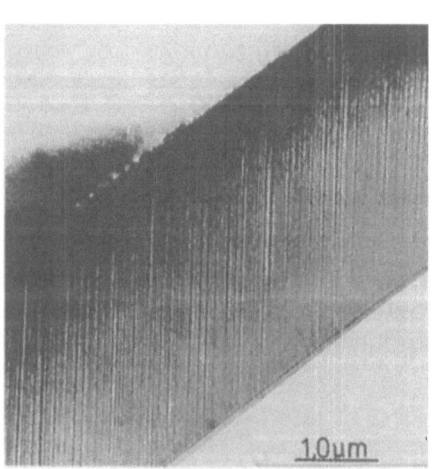

Figure 3: Austenite/martensite microstructure in partly transformed FeNiAl—alloys. Internal twinning inside the martensitic phase α_M, TEM

II   Variables and boundary conditions

Thermal cycling normally comprehends the temperature range between A_f and M_f: $\Delta T > |A_f -M_f|$. There is a variety of possibilities for mechanical cycling. Conventional fatigue takes place in the stable β phase: $T > A_f$. The four other cases may be obtained from figure 1: strain or stress induced $\beta \rightarrow \alpha_M$–transformation above M_s, a $\beta + \alpha_M$–phase mixture between M_s and M_f and fatigue of the completely transformed martensite α_M, i.e. in the pseudoplastic state.

There are different boundary conditions for stresses and strains.

1. $\sigma = 0$ for $\beta \rightarrow \alpha_M$, and $\alpha_M \rightarrow \beta$ represents a free motion during thermal cycling. The additional requirement is ϵ = max for the two–way effect.

2. $\sigma \neq 0$ for $\beta \rightarrow \alpha_M$, or in the α_M–state, but $\sigma = 0$ for $\alpha_M \rightarrow \beta$ is a requirement for the one way effect.

3. $\sigma \neq 0$ for $\beta \rightarrow \alpha_M$, as well as $\alpha_M \rightarrow \beta$ for the superelasticity.

4. $\epsilon = 0$ represents a completely constrained system in which stresses are maximized: for the two way effect this implies that a stress may be created during cooling, which is relaxed during heating.

5. The one–way effect requires $\epsilon \neq 0$ for formation or deformation of the martensitic state. Retransformation $\alpha_M \rightarrow \beta$ will create a stress for $\epsilon = 0$.

6. There are many conditions in shape memory technology for which $\epsilon \neq 0$, $\sigma \neq 0$ is valid. An example is robot grippers which require motion and exertion of a force by the two–way effect.

All of these conditions can be associated with fatigue phenomena, if mechanical and/or thermal cycles are repeatedly applied (Fig. 4).

Table 1:   Fatigue loading conditions of shape memory alloys

T range / mode of loading	$T > A_f \approx T_0$ stable β	$A_f > T > M_s$ pre– martensitic	$M_s > T > M_f$ transform– ing β	$T < M_f$ martensite
mechanical	fatigue in austenite	pseudo– elastic	stress in– duced in $\beta + \alpha$ mixture	fatigue in martensite
thermal	cycling in the range of $\Delta T = T_1 - T_2$; $\quad T_2 > A_f$; $\quad T_1 < M_f$			

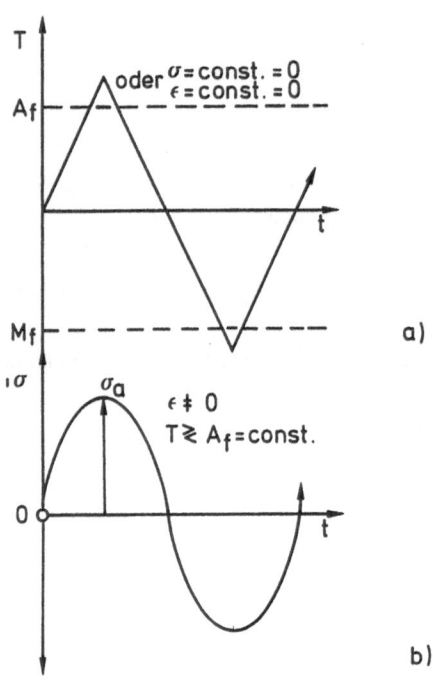

Figure 4: a) Thermal cycling, temperature amplitude, $\Delta T > (A_f - M_s)$; b) mechanical cycling, stress amplitude, σ_a, schematic.

a)

b)

III Experimental observations

Figure 5 provides evidence for the complexity of fatigue in shape memory alloys. Purely thermal cycling ($\sigma = 0$) leads to considerable work hardening of β–CuZn. Its transformation is not associated with a volume change ($\Delta v_{\beta\alpha} \approx 0$). Alloys with a volume change (FeMnX: $\Delta v_{\gamma\epsilon} \approx -0.01$; FeNiX: $\Delta v_{\gamma\alpha} \approx +0.03$) show even more hardening. Minimum hardening is observed for NiTi—base alloys with a high yield stress R_p. This specifies the prerequisite for minimum thermal hardening: $\Delta v_{\beta\alpha} = min$, $R_p = max$, in addition to $\gamma_{\beta\alpha} = max$, i. e. a maximum crystallographic shear strain, for all shape memory alloys. A consequence of the accumulation of the defects which produce the hardening is a characteristic change in the stress–strain–curve of the superelastic state (Fig. 5b). There is a reduction of elastic strain as well as of hysteresis. Repetition of the one–way effect may cause its reduction. This is especially true for the Fe–base alloys, with their transformation volume change.

Mechanical fatigue depends on the temperature level. In summary: crack formation and growth is rapid in the β–state unless grain boundaries are folded and desegregated by a hot–rolling treatment. In the martensitic state cracks form easily, but their propagation is sluggish and consequently life not anomalously short (Fig. 6).[9]

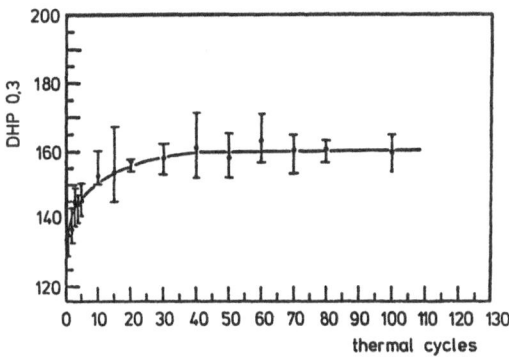

Figure 5: Effects of cycling on properties
of β–CuZn (60.5 Cu, 39.5 Zn wt.%,
M_s = –70)

a) Work hardening, thermal cycling

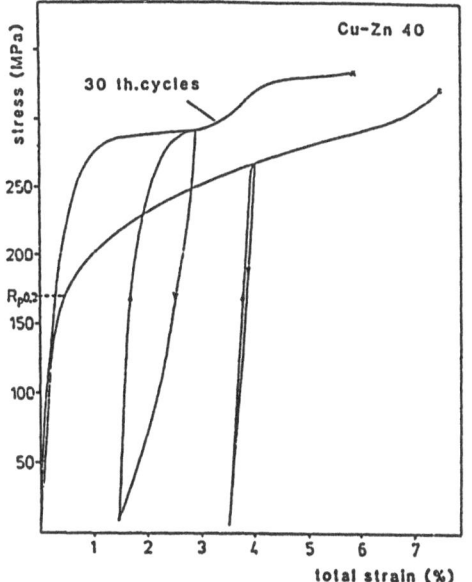

b) Increase of yield stress, hysteresis,
and pseudo–elastic behavior, after
thermal cycling.

Finally, not only mechanical properties do fatigue but also the temperatures and course of the transformation (Fig. 1 and 7). Mechanical cycling of β lowers the transformation temperatures and raises the hysteresis.

IV Microscopic Aspects

Fatigue is due to irreversible structural changes which accumulate during repeated cycling. Special effects in SMA are due to motion of transformation interfaces (α_M/β) and the microstructural disorder in martensite (Fig. 3). In non–transforming alloys dislocations, dislocation groupings, and persistent band structures are the defects which form due to periodic external or thermal

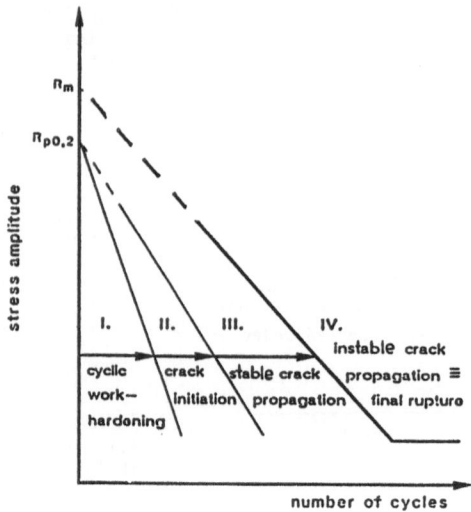

Figure 6a: Stages of development of mechanical failure in a schematic σ–N–curve

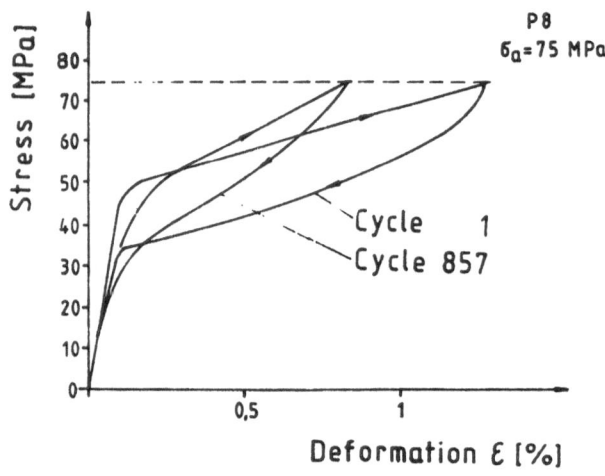

Figure 6b: Effect of mechanical cycles on the shape or the pseudoelastic hysteresis loops (alloy as Fig. 7b)

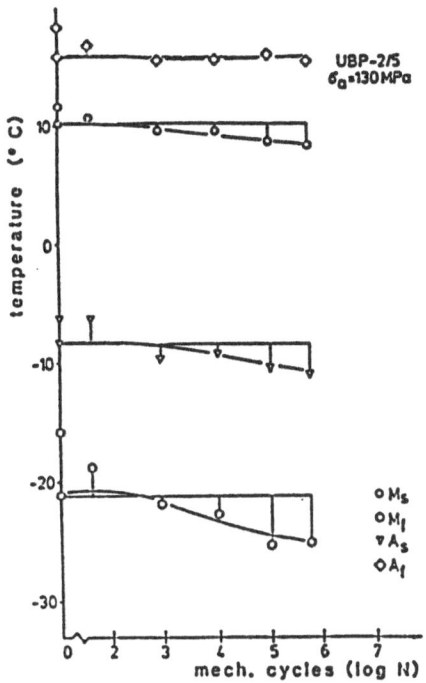

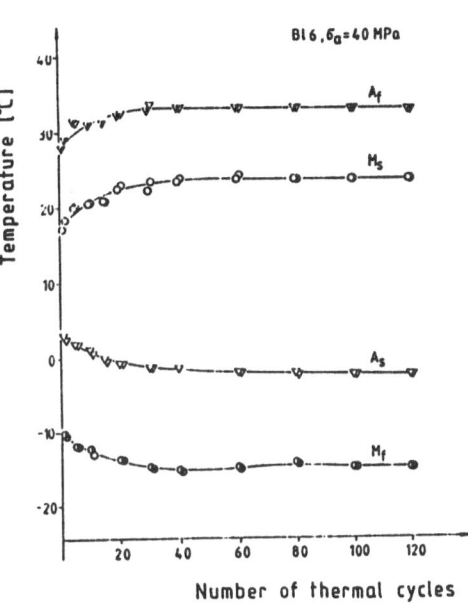

Figure 7a: Mechanical cycling lowers transformation temperatures and increases hysteresis (70 Cu, 26 Zn, 2.9 Al, M_s = + 8), pseudoelastic

Figure 7b: Thermal cycling raises temperatures of martensite formation and lowers that for the start of reversion (72.4 Cu, 24.3 Zn, 3.3 Al, M_s = + 10)

stress. These are enhanced in SMA by incompatibilities associated with shear transformation and amplified further by changes in volume. In all polycrystals the grain boundaries and their close environments are sensitive areas for fatigue effects.

Additional structural changes in SMA are:

1. local zones of disorder in the β–phase, due to local motion of dislocations, starting with the formation of anti–phase–domain boundaries,

2. residual martensite: disorder and lattice defects stabilize this phase against reverse transformation,

3. newly formed interfaces: α_M/β–phase boundaries, twin– and varient boundaries inside martensite.

Figure 8 provides evidence for some of the structural changes. Dislocations found in the β–phase may be of particular nature (i. e. Burgers vector) if they had formed in martensite and re-transformed. Intramartensitic sliding along twin boundaries and faults seems to be easy and the cause for characteristic tongue shaped extrusions. Martensite phase– and variant–boundaries limit the extend of this slip but act as sites for the initiation of fatigue cracks. Other sites for crack initiation are $\beta\beta$–grain boundaries or intersections of localized bands of transcrystalline strain as in non–SMA–alloys. The observed new mechanism explain the bulk effects. Crack formation is

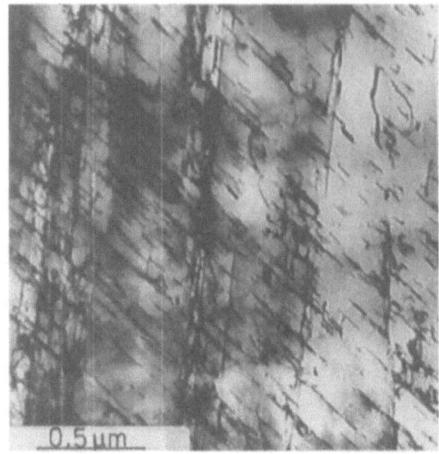

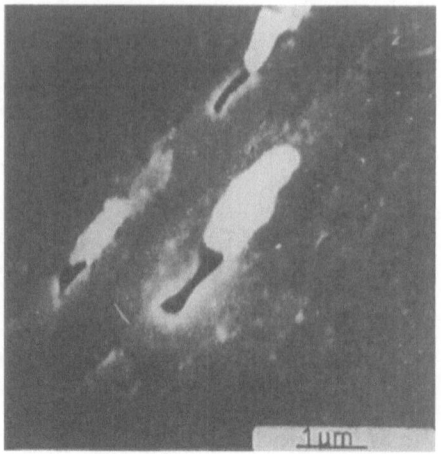

Figure 8a: Examples for microstructural changes during fatigue of Cu–base SMA. Dislocations left in the β–phase, TEM (alloy as Fig. 5a)

Figure 8b: Extrusions formed during mechanical fatigue in the pseudoelastic state (alloy as Fig. 7b)

Figure 8c: Cracked $\alpha^+\alpha^-$–boundaries in the completely transformed condition inducing multiple crack growth (70.6 Cu, 26.8 Zn, 3.6 Al, M_s = + 120 °C)

extremely frequent in martensite and a rare event in the β–phase. This effect has a favourable effect on bulk fatigue life of the alloy in martensitic conditions, which is not plausible on the first sight, but explainable by the observations of multiple crack growth.

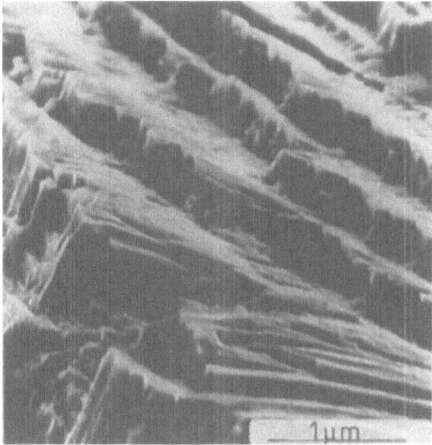

<u>Figure 9a:</u> Fractal fracture surface of embrittled and fatigued martensitic alloys

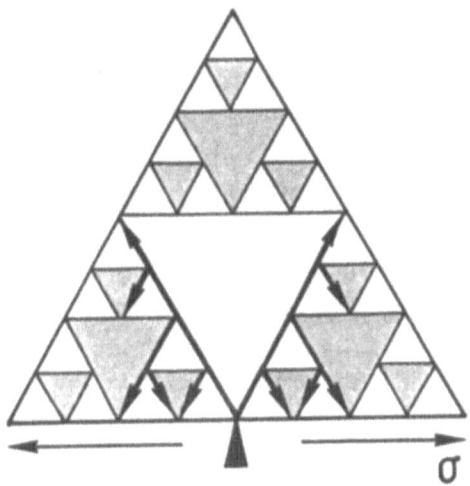

<u>Figure 9b:</u> Formation of fractal fracture, schematic

V   Fractal and non–fractal Fracture

The final stage of fatigue life is rupture (Fig. 9). If rupture occurs as the final stage of a fatigue test or tensile test a very rugged fracture surface is found for partially or completely martensitic alloys. Fracture in the β–state is intercrystalline, on {100}–cleavage planes, or {110}–slip planes.[9] An analysis of fracture surfaces provides evidence for a fractal structure of the fracture of martensite. Fracture in the β–phase is non–fractal. Especially intercrystalline fracture is not at all self–similar, while multiple cracking takes place in martensite especially after fatigue loading or after embrittlement of the intra–martensitic interfaces.[12] Induction of multiple crack branching may have a favourable effect on final fracture toughness. On the other hand it has to be noted that the subcritical changes in microstructure cannot be ignored because they will strongly affect the shape memory properties, i. e. amount of trained–in two way memory, transformation temperatures, amount of transformation, course of transformation.

VI Summary and conclusion

The origin of fatigue in shape memory alloys are structural changes which can be induced either by external cyclic loads or temperature cycling through the transfomation range between T_1 > A_f and T_2 < M_f. Transformation shears and volume changes induce stresses in addition to those known from non–SMA–alloys. As a consequence of the martensitic transformation defects other than normal dislocations and grain boundaries play a role in shape memory alloys:

a) dislocations that have undergone a phase transformation,

b) zones of disorder and antiphase domain boundaries in β,

c) residual martensite in β,

d) α–β interfaces,

e) martensite plate boundaries, twin boundaries inside α_M.

There are two aspects of fatigue which are related to these structural defects:

1. Mechanical fatigue:

Cyclic hardening, formation of surface phenomena, (tongue–like extrusions–intrusions), crack initiation, crack propagation, final rupture. The martensitic structure is capable to multiple crack formation. Transforming β may produce retarded crack growth by transformation localized at the crack tip. These effects explain the observation that fatigue life of the transforming alloys is not worse as compared to normal non–transforming ones.

2. Shape memory fatigue:

These are the changes in the transformation behavior including the extend of the one–way and two–way effects. A dislocation forest in β will retard the transformation i. e. lower M_s and raise the hysteresis. Local dislocation groupings, as sites of internal stresses will favour the $\beta \rightarrow \alpha_M$ transformation, i. e. raise M_s. The same is true for small zones of residual martensite and disordered austenite. These structures form during thermal cycling and will raise the start of martensitic transformation.

The structural features listed as a)–e) are intentionally introduced during training. They are then responsible for the amount of the two–way effect. Mechanical fatigue is always related to shape memory fatigue in a complex way. Both can be understood by a subtle analysis of the microstructural details.

Acknowledgement

This work was supported by the German Ministry of Technology and Research (BMFT O3M 5006 B5). Experimental results from the following former co–workers were mentioned in this paper: Dr. M. Thumann, Dr. M. Sade, J. Kumpfert.

References

1) E. Hornbogen, Microstructure and Mechanisms of Fracture. 6th International Conference on Strength of Metals and Alloys, Melbourne, Australia, p. 1059–1068 (1982)

2) K. N. Melton and O. Mercier, Fatigue Life of Cu–Zn–Al Alloys. *Scripta Met.*, **13**, 73–75 (1979)

3) J. Perkins and W. E. Muesing, Martensitic Transformation Cycling Effects in Cu–Zn–Al Shape Memory Alloys. *Met. Trans.*, **14A**, 33–36 (1983)

4) J. Perkins and R. O. Sponholz, Stress–Induced Martensitic Transformation Cycling and Two–Way Shape Memory Training in Cu–Zn–Al Alloys. *Met. Trans.*, **15A**, 313–320 (1984)

5) K. Shimizu, Aging and Thermal Cycling Effects in Shape Memory Alloys. *J. Electron Microscopy*, **34**, 277–288 (1985)

6) M. Sade, K. Halter, E. Hornbogen, The Effect of Thermal Cycling on the Transformation Behavior of Fe–Mn–Si–Shape Memory Alloys, *Z. Metallkde.*, **79**, 487–491 (1988)

7) M. Sade and E. Hornbogen, Fatigue of Single and Polycrystalline Cu–Zn–base Shape Memory Alloys, *Z. Metallkde.*, **79**, 782–787 (1988)

8) M. Sade, J. Kumpfert and E. Hornbogen, Thermomechanical and Pseudoelastic Fatigue of a Polycrystalline CuZn24Al13 Alloy, *Z. Metallkde.*, **79**, 678–683 (1988)

9) M. Thumann and E. Hornbogen, Thermal and Mechanical Fatigue in Cu–base Shape Memory Alloys, *Z. Metallkde.*, **79**, 119–126 (1988)

10) M. Sade, E. Cesari and E. Hornbogen, Calorimetry of CuZnAl–Single Crystals during Pseudoelastic Transformation, *J. Mat. Sci.–Lett.*, **8**, 191–193 (1989)

11) E. Hornbogen, Fatigue of Cu–base Shape Memory Alloys. In: *Engineering Aspects of Shape Memory Alloys.* T. Duerig, Ed., Butterworth, London/Boston (1990) (to be published)

12) E. Hornbogen and N. Jost, Eds., The Martensitic Transformation in Science and Technology. DGM Informationsgesellschaft, Oberursel, FRG (1989)

13) E. Hornbogen, Fractals in Microstructure of Metals. *Int. Mat. Rev.*, **34** (to be published) (1989)

Matrix Cracking of Cross-ply Laminates under T-fatigue and Thermal Loading

H. Eggers, H.C. Goetting, W. Hartung, H. Twardy
Institut für Strukturmechanik, DLR-Braunschweig
Flughafen, D-3300 Braunschweig

Abstract: The present study concentrates on resin cracks in UD-layers of cross-ply laminates. The material T300 / 914C was tested mainly under T-fatigue, thermal cycling at space conditions and overheating, in order to analyse the damage mechanisms for resin layer cracks and their causes.

1. Introduction

Because of the complicated texture and structure of laminates, different kinds of damage appear simultaneously, inducing and magnifying each other. Most severe is the synergism between delaminations and layer cracking, especially for T-C-fatigue. The closely spaced matrix cracks damage the interfaces between the layers and accelerate the delamination growth. Generally, the local buckling of the separated sublaminate initiate the failure of the component. An indicator for the mentioned synergism is the large difference between the energy release rate (ERR) at the delamination front for cracked and uncracked layers, Figure 1.

As a first step the matrix cracking of the imbedded layers will be studied separately. The tests, limited to cross-ply laminates, are accomplished for T-fatigue, thermal cycling under space conditions and overheating, which may happen accidently, Table 1.

Material	Stack	Test conditions	Test parameter	Measured values
T300 / 914C	$[\,0_s\,/\,90_m\,/\,0_n\,]$	T-fatigue $R = 0.1$	$n = 1{,}2 \quad m = 1{,}2{,}3{,}4$ $\sigma_{g^*} = 1050 - 1440\ \text{N/mm}^2$ $N = 5\,000 - 4\,000\,000$	Crack distance $\sigma_{Ult}^{cl},\ E_1,\ G_1$
T300 / 914C	$[\,0_s\,/\,90_m\,/\,0_n\,]$	Thermal cycling	$n = 1{,}2 \quad m = 1{,}2{,}3{,}4$ $-160\ ^\circ\text{C} \le T \le 100\ ^\circ\text{C}$ $N \le 2000$	Crack distance
Different materials	$[\,\pm 45\,]_{2s}$	Thermal cycling	$-160\ ^\circ\text{C} \le T \le 100\ ^\circ\text{C}$ $N \le 3500$ Vacuum N_2-atmosphere Cycle time	Crack distance $\sigma\text{-}\varepsilon$-diagram. $\sigma_{Ult.}$ E_*
T300 / 914C	$[\,\pm 45\,]_{2s}$	Overheating	$T = 180,\ 230,\ 260\ ^\circ\text{C}$ N_2-atmosphere	Mass loss $\sigma\text{-}\varepsilon$-diagram. $\sigma_{Ult.}$ E_*

Table 1. Test configurations for cross-ply laminates

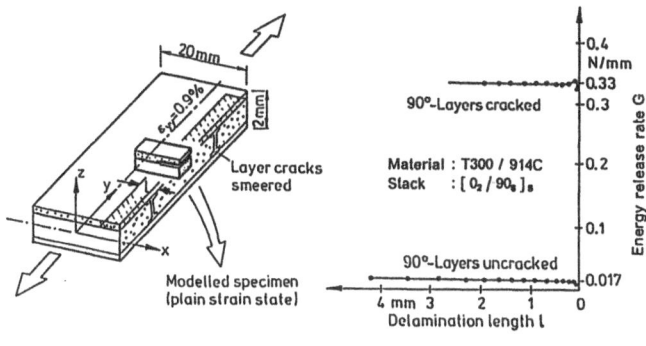

Figure 1.

Influence of layer cracks on the energy release rate at the delamination tip (Evaluated by E. Haug)

2. T-Fatigue loading

In MD-laminates the UD-layers crack in parallel to the fibers under tension stresses. The crack density increases with the number of load cycles, until a saturated state, the characteristic damage state (CDS), is achieved, Figure 2. In a a-log N-scale the crack distances a decrease linearly with the number of load cycles N, until the mean crack distance for the CDS is reached. In Figure 3 the measured crack distances are plotted for a $[0_1 / 90_4 / 0_1]$-laminate. For each 90°-layer removed from the stack the crack distances decrease by $\Delta a(N) \approx 0.15$ mm. For a $< a_{CDS}$ the slope of the lines decrease with the load level and with the number of layers. In order to find the limit values for extremely stiff top layers tests are in preparation with $[0_4 / 90_2 / 0_4]$- and $[0_4 / 90_4 / 0_4]$-stacks. For a stack of ≤4 UD-layers the CDS depends solely on the thickness of the UD-stack and is independent from the load level and the stiffnesses of adjacent sublaminates.

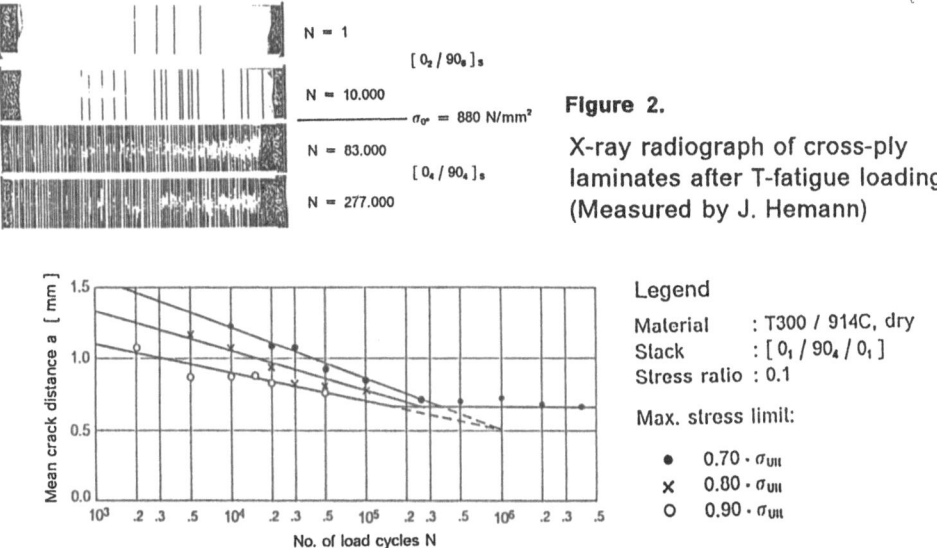

Figure 2.

X-ray radiograph of cross-ply laminates after T-fatigue loading (Measured by J. Hemann)

Legend

Material	: T300 / 914C, dry
Stack	: $[0_1 / 90_4 / 0_1]$
Stress ratio	: 0.1

Max. stress limit:

●	$0.70 \cdot \sigma_{Ult}$
×	$0.80 \cdot \sigma_{Ult}$
○	$0.90 \cdot \sigma_{Ult}$

Figure 3. Crack distances versus load cycles

For various load cycles the frequencies of the crack distances can be described by a two-parametric Weibull-distribution, which changes gradually to a symmetric Gauß-distribution for the CDS, Figure 4. It is typical for the saturated state, that the extreme crack distance is about twice as large as the mean crack distance measured most frequent.

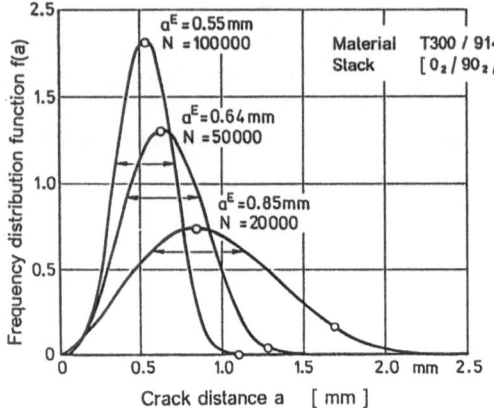

Figure 4.

Weibull-frequency versus crack distances for different load cycles

In order to study the crack pattern more precisely, adjacent crack distances a_1, a_2 are marked by a dot in a a_1-a_2-plane, which is subdivided in squares of 0.05 mm pixelsize. In each square the dots are summed and lines of equal frequencies are plotted for the cluster sums, Figure 5. If a third crack is forming between two existing cracks, the sum of $a_1 + a_2$ is constant. This holds for the dashed line in Figure 5, along which the frequencies follow a Gauß distribution. For small crack distances, where the stresses close to both crack surfaces are interdependent, a new crack is likely to form in the middle ($a_1 = a_2$). Moreover, crack distances below 0.35 mm will not subdivid any more and crack distances below 0.15 mm will not be generated.

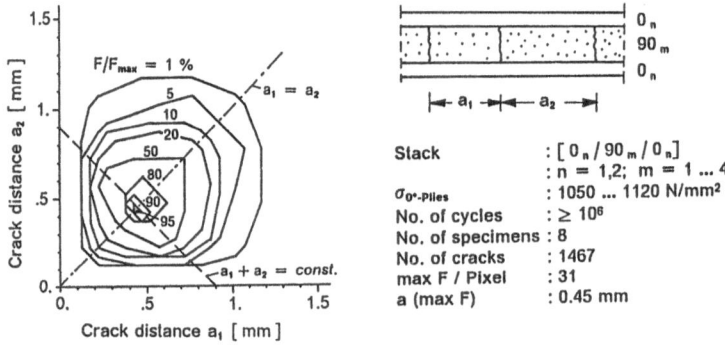

Figure 5.   Envelope lines for the frequencies of adjacent crack distances

The results depicted in Figure 5 verify, that in cross-ply laminates layer cracks will be initiated predominantly by the extreme inplane tensile stresses. Figure 6 depicts these stresses in the middle of a 90°-layer block for various crack distances. The tensile stresses close to the interface are nearly constant between adjacent cracks.

Even though these stresses overstep the ultimate strength of $\sigma_{ult} \approx 50 \ N/mm^2$ significantly, a layer crack generally does not form. Voids, which have the potential to
progress under high stresses, cannot open freely in the vicinity of the 0°-layers and
therefore will not grow. This conclusion is supported by the decrease of ERR, when
a crack starting in the middle of a 90°-layer steps close to the interface, Figure 6.

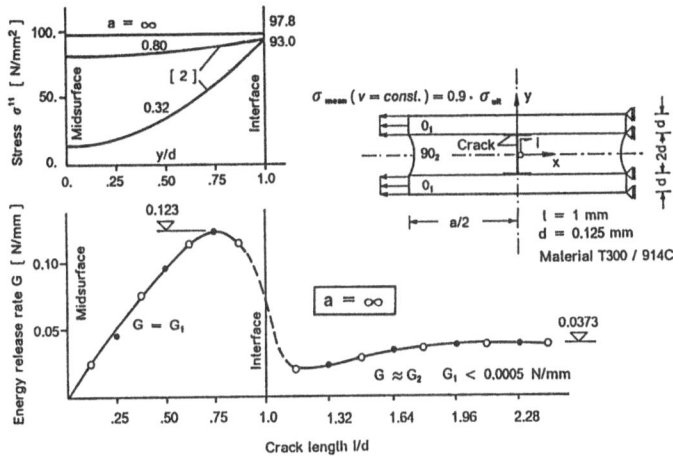

Figure 6. Tensile stresses in the mid-cross-section between two cracks
and ERR after crack initiation at the midpoint

The cracks will probably be initiated in the inner zone of the 90°-layers. Once a crack
is formed there, it will accelerate due to the stress increase towards the interface.
This assumption is supported by the fact that partly cracked layers were never
observed by tests.

When a crack reaches the interface, the sudden contraction of the crack surface
causes extremely high peeling stresses at the interface. They dismantle and break
some fibers and generate microdelaminations, Figure 7.

The described results give an insight into the generation of layer cracks. Tests are
in preparation to measure the stiffness and strength of damaged 90°-layers strained
by interlaminar stress components. Hopefully, these tests combined with an analysis
of the inplane stresses can be used to determine damage conditions for layer
cracking.

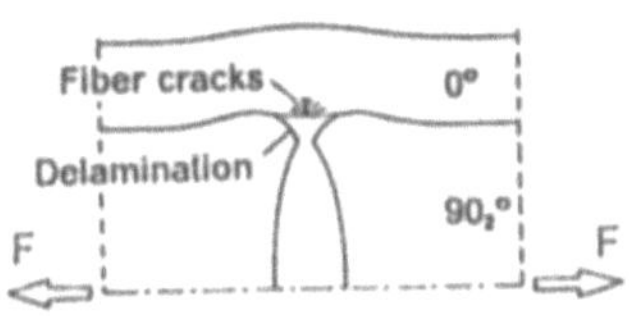

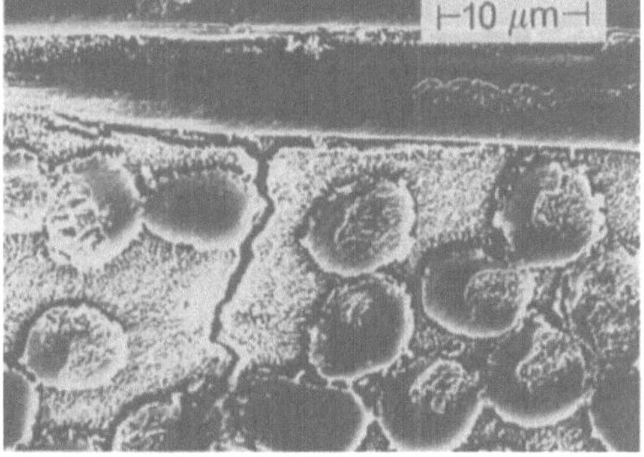

Figure 7. SEM-picture of a layer crack
with microdelaminations
(Measured by H.J. Seifert)

Summarized results for T-fatigue tests on T300 /914C material:

- Crack distances decrease linearily in a a-log N-diagramme
- CDS depends on the layer thickness and not on the load level
- Crack distances ≤0.15 mm were not measured
- Crack distances ≤0.35 mm will not subdivide
- Inplane tensile stresses normal to the fibers are dominant for resin layer cracks
- Layer cracks form suddenly and often cause microdelaminations
- Layer cracks damage the top fibers in adjacent off-axis layers
- The fiber cracking of a total layer follows adjacent resin layer cracks

3. Thermal Cycling

Space structures placed in low earth orbits may experience up to 50.000 thermal cycles during their service life. Therefore, different materials were tested under space conditions, [1]. The tests, depicted in Table 2, were conducted in vacuum, where the heating to +100 °C and the cooling to -160 °C was effected by infrared radiation and by absorbtion of radiation by cold surfaces, respectively. The chosen heating and cooling rates correspond to realistic space conditions, but the hold time at both temperature extremes was sharply reduced on the supposition that no damage develops at constant temperature.

Stacking sequence: $[\pm 45°]_{2S}$, $[\pm 45°]_{4S}$
Material:

Laminate	Resin	Curing temp. °C
Compimide 65 FWR/JM6	Bismaleimide	210
Fibredux. 914C−TS−5	Epoxy	175
LY556/HT 976/JM6	Epoxy	175
Hercules 4502/JM6	Epoxy	180
Carboform SO 60/92/51	Epoxy	120

Test spefications:

Medium	Cycle time min	Cooling rate K/min
Vacuum	62	4.7
N$_2$	52	
	23	32.5

Temperature range: −160°C to +100°C
Total no. of cycles: ≤ 3500

Measurements:

Measurements	No. of cycles	Test
Stiffness	10,50,150,400, 800,1500	Three point bending test, max ε ≤ 2‰
Strength	400,800,1500	Tensile test

Specemien size: 195∗65 mm

Table 2.

Test configurations for thermally cycled laminates

Tests in vacuum are costly and time consuming (cycle time 62 min). Therefore, they were compared with tests in N$_2$-atmosphere, conducted at similar thermal condtions but reduced cycle times and higher cooling rates, Figure 8. In both environments the degradation of strength, stiffness and crack density were very similar for equal cycling modes. Only an increase of the cooling rate at short cycle time increases the damage. Therefore, space-rated facilities are not required for degradation tests at thermal cycling.

Figure 9 depicts the acoustic energy, emitted at two individual thermal cycles. The highest damage rate were found during the first few cycles. When the cycling continued the damage increased further but at a diminishing rate. The onset of damage at a certain temperature in the course of a cycle seems to be typical for composite materials.

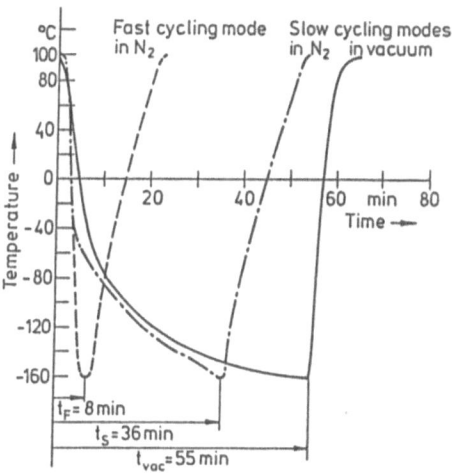

Figure 8.

Temperature versus time courses for different cycling modes

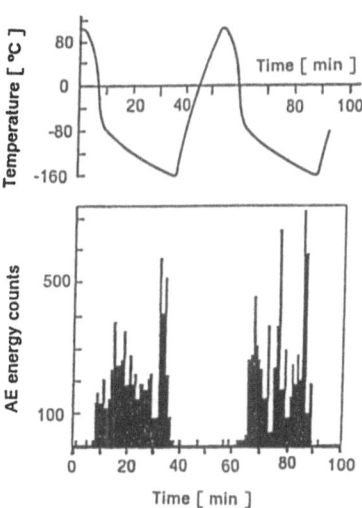

Figure 9.

Acoustic energy emitted in a Comprimide 65 FWR-laminate

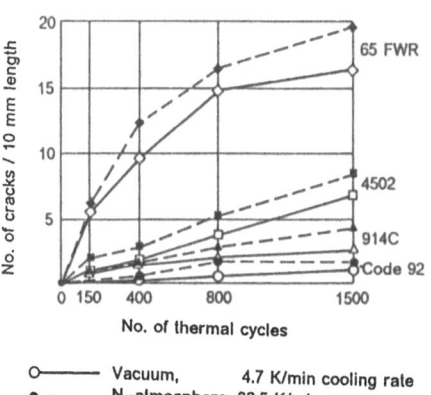

○——— Vacuum, 4.7 K/min cooling rate
●----- N₂-atmosphere, 32.5 K/min

Figure 10.

Crack density due to different cycling modes

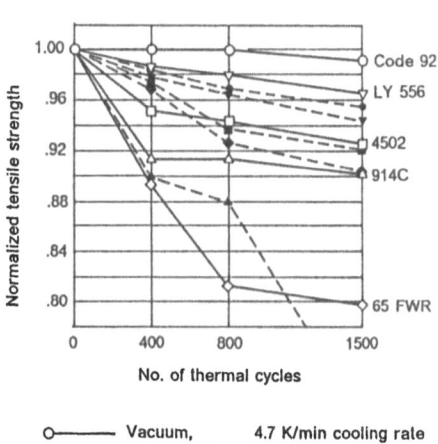

○——— Vacuum, 4.7 K/min cooling rate
●----- N₂-atmosphere, 32.5 K/min

Figure 11.

Residual tensile strength after thermal cycling in different environments

In Figure 10 the number of cracks per 10 mm ply length are plotted versus the number oft termal cycles. For all of the materials considered damage is most severe at fast cycling. Therefore, a critical damage state will be achieved faster at a short cycling mode than in a slow mode. This may be of some importance for the qualification testing of space structures.

Figures 11 and 12 show respectively the degradations of tensile strength and stiffness due to thermal cycling. For slow cycling in vacuum the stiffness degradation is ≤2 % smaler than the values plotted in Figure 12 for fast cycling in N_2-atmosphere. These curves are not plotted, in order to avoid line jams.

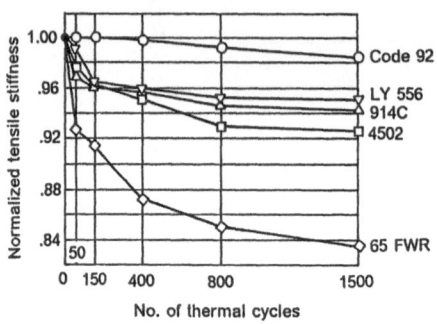

N₂-atmosphere, 32.5 K/min cooling rate

Figure 12. Residual stiffness after thermal cycling

Summarized results for thermal cycling on different materials:

- Slow cycling generates less damage than fast cycling
- Damage increase for T ≤ -80 °C
- Damage progress with the number of thermal cycles
- Damage still progress after 3500 thermal cycles
- Damage received in vavuum and in N_2-atmosphere depends on the cooling rate only.

4. Overheating

Structures fabricated by carbonfiber reinforced epoxy can be exposed to temperatures up to about 40 °C below the softening point without reducing their strength permanently. But it may happen accidently, that a laminated structure is overheated locally beyond that temperature. The questions arise, to what extent is the structure damaged and how can it be detected?

For the following tests laminates with a stacking sequence of $[\pm 45]_{2s}$ were used, fabricated of T300 / 914C. The dry resin has a softening point at $T_{GA} \approx 180$ °C and the glass point were measured at $T_{GB} = 208\,°C$. In N_2-atmosphere the specimens were exposed to elevated temperatures of 180, 230 and 260 °C, until a mass loss of 0.8, 2.0 and 4.0 % were detected. Then the specimens were cooled down to ambient temperature and the stiffness, strength and stress-strain-relations were measured in air-atmosphere.

Figure 13 shows the mass loss with respect to the storage time for different temperature levels. If the specimen is exposed to temperatures lower than 40 °C below the softening point, the mass loss can be neglected. But it increases rapidly, when the temperature level oversteps the glass point. Due to ESA-regulations for space crafts, the mass loss is limited to 0.2 % after 5.5 h exposure time, in order to avoid sedimentations on lenses or electronic equipment.

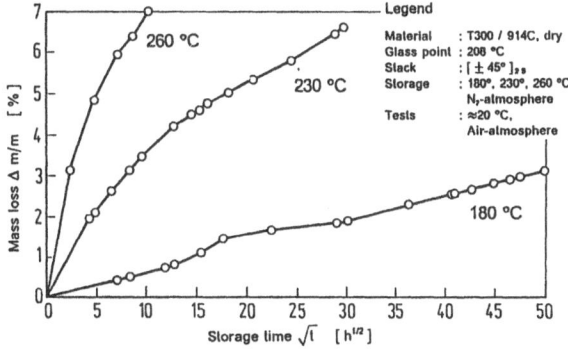

Figure 13.   Mass loss in 914C-material due to elevated temperatures

For temperatures lower than 40 °C below the softening point the strength reduction is negligible, but it increases rapidly for temperatures beyond the glass point, Figure 14. On each curve in Figure 14 the dots mark the intended mass loss of 0.8, 2.0 and 4.0 %, for which the storage time and the residual strength were measured.

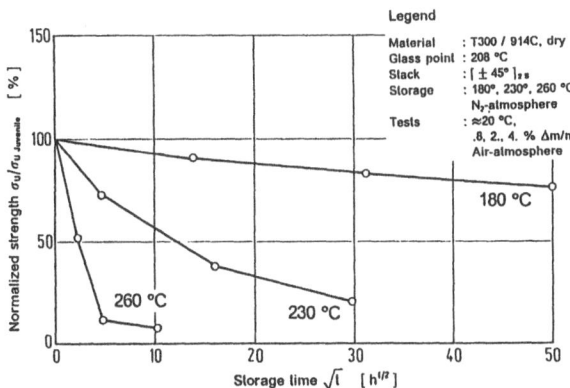

Figure 14.   Residual strength after storage at elevated temperatures

The elasticity properties do not change much with thermal exposure, Figure 15. Apparently, the increase of the residual stiffness caused by the postcuring process of the resin is balanced by chemical degradation, which is predominant only for long exposure times. Figure 16 depicts the σ-ε-diagrammes for juvenile specimens. These diagrammes hold also for specimens exposed to temperatures beyond the glass point with the exception, that the diagramme is broken at the limit stress depicted in Figure 14.

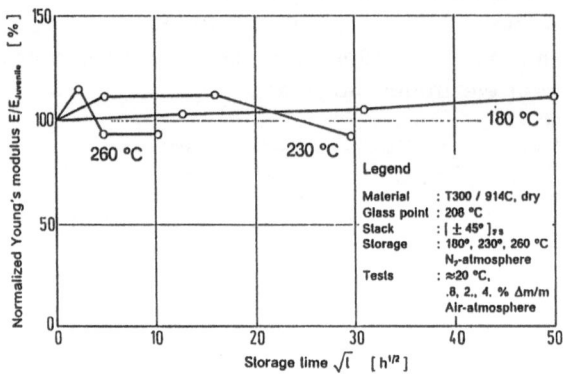

Figure 15.   Young's modulus after storage at elevated temperatures

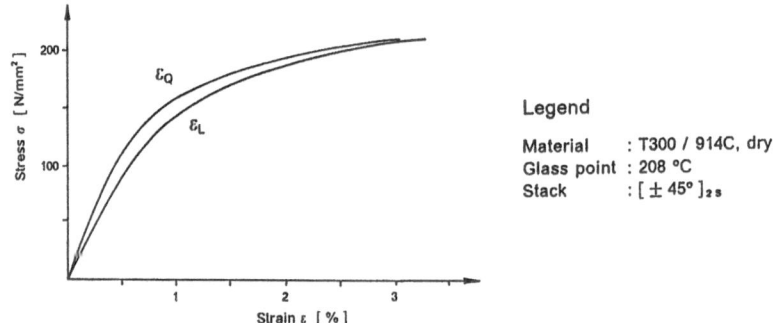

Figure 16.   Stress-strain-diagrammes of juvenile angle-ply specimens

Summarized results on overheating tests at T300 / 914C material:

- Mass loss increases for $T > T_{GB}$
- Residual strength decreases rapidly for $T \gg T_{GB}$
- Elasticity properties and the finishing of the structure remains unchanged
- Even for $T > T_{GA}$ layer cracks form and dense
- Overheating reduces the strength signifficiantly, but it cannot be detected by visual inspection or by nondestructive vibration tests.

5. Closing Remarks

The present study, partly supported by ESA-contracts, is not finished and will be extended. Additional tests for thermal cycling and at very low temperatures are in preparation, in order to find limit values for the CDS.

6. Bibliography

[1] H. W. Bergmann et al.: *Mechanical Properties and Damage Mechnisms of Carbonfiber-Reinforced Composites - Compression Loading*. DFVLR-FB 88-41, 1988, 262-293.

[2] A. Höhn: *Schichtrißbildung in Kreuzschichtlaminaten*. Study work , T.U. Braunschweig, 1989.

LOAD AND DAMAGE DEPENDENT THERMAL EFFECTS IN CFRP-LAMINATES

Karl Schulte, DLR, Institut für Werkstoff-Forschung,
5000 Köln 90, W.-Germany

Harald Neubert, Sintermetallwerk Krebsöge, 5609 Hückeswagen,
W.-Germany

H. Harig, Faserinstitut Bremen, 2800 Bremen 1, W.-Germany

1.   Introduction

The temperature variation during monotonic or cyclic loading of cfrp-
laminates is generally caused by the laminate stress state and the
energy dissipation during deformation work. Local stress variations
may result from changes of the external load or from the formation of
cracks, delamination or fibre fracture in the stressed volume. Due to
the thermoelastic effect there exists in case of an elastic deformation a linear relation between stress and temperature in the fibres
as well as in the matrix. This means, that always a mean value of
temperature change is measured, because material constants differ.
The dissipation of deformation work is mainly due to viscoelastic and
plastic deformation of the matrix. It leads to an increase in
temperature.
Of special interest are the temperature changes during cyclic loading. The mean temperature value of the specimen characterizes the
area of the stress strain hysteresis loop. Changes of this temperature indicate unstable deformation processes which means that damage
development occurs.
On the other hand, changes of the temperature oscillation due to the
thermoplastic effect indicate changes of the stress amplitude which
also is a result of an increasing damage development resulting in
stress redistributions.

Thus, there seem to exist two sensitive methods of temperature measurement which can be used to evaluate damage development under monotonic and especially under cyclic loading. It is the aime of this

paper to show that temperature measurements can often be more sensitive than strain measurements but if they are performed together, they benefit from each other and give additional informations.

2.   Experimental

All tests were performed on continuous carbon fibre reinforced epoxy matrix systems (PAN-based T 300 carbon fibres from Toray or HTA from Toho/LY 556/MY 720 epoxy matrix resins from Ciba-Geigy). Two types of laminate stacking sequences, unidirectional $[0_8]$ and cross-ply $[0_2,90_2,0_2,90_2]_s$ were investigated. The tests were performed on servohydraulic testing machines. The specimen temperature was measured with thermocouples (Fe-CuNi) which were directly glued on the specimen surface. To avoid heat flux from the actuator into a specimen, the grip temperature was hold constant by a special cooling system. To compare the test results with established methods, the elongation of the specimen was continuously monitored, allowing to calculate the secant modulus. In addition, in selected specimens, the variation of the electrical resistivity due to damage development was continuously measured [1].

3.   Results and discussion

Composite materials mainly fibre reinforced plastics, are gaining increasing importance for structural design. For a reliable use their durability is of essential importance.
In metallic materials failure occurs by crack initiation and subsequent propagation of a single crack. However, in composite materials damage is associated to a simultanious initiation and multiplication of interfibre cracks, delaminations and fibre breaks [2, 3] which all form prior to final rupture. The in-situ investigation of damage development during loading remains comparitively difficult. While in metals crack length can be taken as a damage analogue, in composites such a simple value can not be chosen.

In the past, stiffness reduction, continuously registered during loading, turned out to be an appropriate damage analogue [4]. As polymere materials show an internal heating due to cyclic loading a temperature variation can as well be used to monitor damage [5-8]. During fatigue loading of a composite material the heat generation can mainly be related to the matrix material used, its viscoelastic behaviour, internal fraction and the amount of crosslinking. The type of fibre, its volume fraction and orientation and the stacking sequence in a composite laminate additionally effect matrix loading and therefore heat generation. In addition to the materials aspects, conditions as shape of sample, test frequency and load level influence the amount of heating. However, also under static loading conditions a temperature variation of a material can occur. In 1851 Thomson (Lord Kelvin) [9, 10] showed the proportionality between the load change applied and the resulting temperature variation for an isotropic material under adiabatic elastic deformation and uniaxial stress, where

$$\Delta T = -\alpha_1 \, T_0 \, \Delta\sigma \, / \, c \, p$$

α_1	K^{-1}	= linear coefficient of thermal expansion	
T_0	K	= ambient temperature	
c	$kJ \; K^{-1} \; m^{-3}$	= specific heat capacity	(1)
p	$g \; cm^{-3}$	= specific density	
$\Delta\sigma$	Nmm^{-2}	= stress change	
ΔT	K	= temperature change	

The above equation describes an effect, the so called "Thermoelastic Effect". The amount of temperature change of a certain volume element depends on the change of the sum of principal stresses, on its material factors and on the ambient temperature. For an isotropic material with a positive coefficient of thermal expansion uniaxial tensile stresses lead to a decrease in temperature. The application of the thermoelastic effect on carbon fibre reinforced plastics has first bo performed by Neubert et al. [7, 8].

3.1 Thermoelastic Effect

Fig. 1 is a tension test of unidirectional laminate. Both longitudinal and transverse stresses are plotted together with the temperature change versus the applied stress. The nearly linear increase in strain is accompanied by a steady increase in temperature.

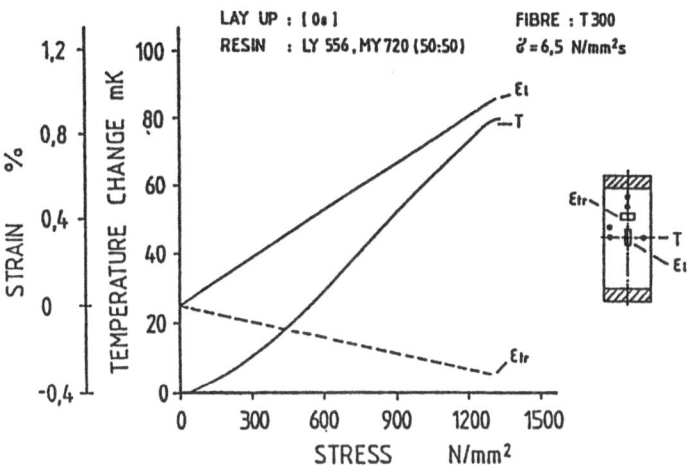

Fig. 1:   Tensile test of a unidirectional laminate. Variation of
          strain and specimen temperature.

The temperature increase is expected because the carbon fibres have a
negative coefficient of thermal expansion into the longitudinal
direction and they therefore heat up under tension load, due to the
thermoelastic effect. This is caused by the dominant influence of the
fibres. (The matrix, however, would cool down because of its positive
thermal expansion coefficient).

In Fig. 2 are shown the results from a cross-ply laminate, containing
0°- and 90°-plies. The fibres in the 90°-plies hinder transverse deformation. They are under compressive load and their temperature change
is negative. In addition, the matrix in the 90°-plies is under tensile loading, leading for the matrix, which has a positive thermal
expansion coefficient, to a cooling down.

Both these effects superimpose the temperature increase in the 0°-
plies, and the result is a cooling down of the intire specimen. At
higher stresses, with the initiation of transverse cracks [2] in the
90°-plies, the stresses in the 90°-plies are locally released and a
load transfer from the 90°-plies into the neighbouring 0°-plies
occurs. With increasing number of cracks the temperature increase due

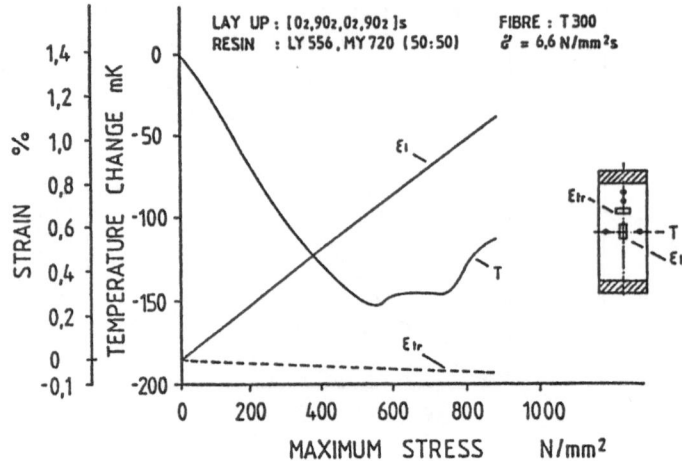

Fig. 2:   Tensile test of a cross-ply laminate. Variation of strain
           and specimen temperature.

to loading of the 0°-plies becomes more and more dominant (the curve
of temperature change goes through a minimum) and at high stresses a
temperature increase can be measured. Viscoelastic deformations at
the transverse crack tips possibly contribute to the temperature
increase.

3.2 Effect of Thermal Heating

Under cyclic loading conditions, which were performed at frequencies
of 10 Hz and with an R-value of R = 0,1 (R = σ_u/σ_o; σ_u = minimum and
σ_o = maximum stress in each load cycle) an internal generation of
heat, due to viscoelastic behaviour and internal friction leads to an
increase in specimen temperature. In Fig. 3 is shown the temperature
development in a cross-ply test piece. A temperature increase due to
dissipation can be observed at the beginning of the test. Later the
temperature remains constant. This has two reasons:
-    Each specimen has, depending on load level, frequency, stacking
     sequence and constituents a stable temperature level [5].
-    A damaged specimen has, if a constant damage state is reached, a
     higher stable temperature level as an undamaged one [5].

For the present test, where the result is shown in Fig. 3, a constant
damage state means, that in early load history (during the first load
cycles) transverse cracks initiate in the 90°-plies, which are also
responsible for the slight stiffness reduction observed at the beginning of the test. Later the number of transverse cracks, the stiffness reduction and the specimen temperature remain constant.

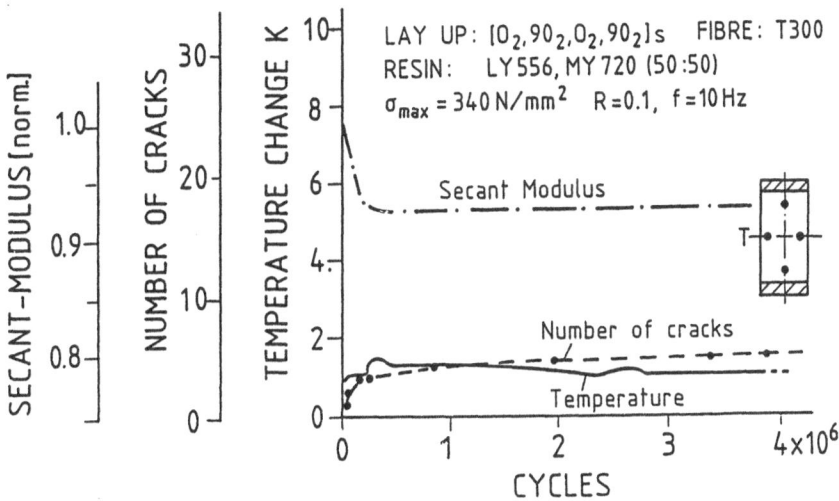

Fig. 3: Fatigue test of a cross-ply laminate at a low load level.
Stiffness reduction, development of transverse cracks and
temperature change is plotted versus the number of load
cycles. Specimen did not fail. Number of transverse cracks
taken from x-ray radiographs.

In the case of continuous damage growth, which will be observed at
higher fatigue load levels, also a continuous increase in specimen
temperature is measured. Fig. 4a shows the temperature change for
four different tests. All tests were performed until specimen rupture
occured. A continuous temperature increase can be observed throughout
the tests. This corresponds with the parallel observation of the transverse crack development (Fig. 4b), where the number of cracks increases with increasing load cycles, and, the higher the cyclic load
level the more transverse cracks are formed.

184

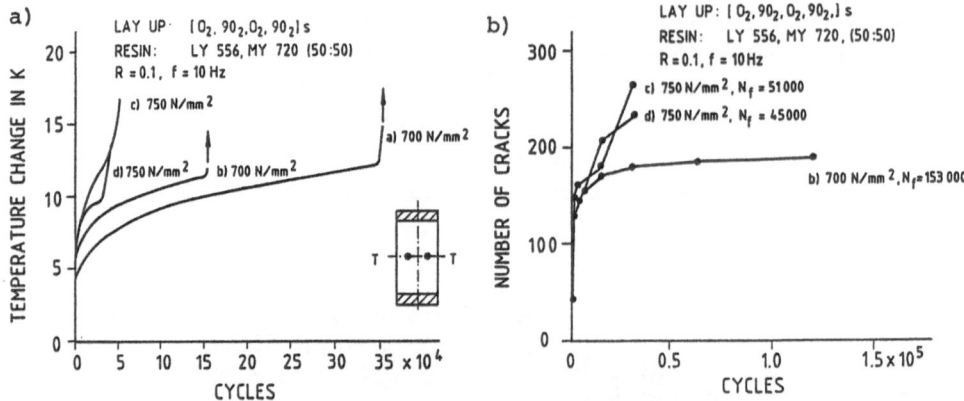

Fig. 4:   Fatigue test of cross-ply laminates at high load levels.
All specimen run to failure.
a)   Temperature change versus number of load cycles.
b)   Development of transverse cracks versus number of
load cycles. Number of transverse cracks taken from
x-ray radiographs.

3.3   Further consequences of temperature measurement

We have seen in the previous sections, that under fatigue loading,
the temperature in a test coupon can not be regarded as constant;   at
test frequencies of 10 Hz, for example, significant increases in temperature have been observed. The variation of the specimen temperature can be used as a means of detecting damage in a composite specimen. Increasing specimen temperature is a result of increasing (matrix) damage.

In Fig. 5 is shown the result of a fatigue test on a $[0_2,90_2,0_2,90_2]_s$
cross-ply laminate [1]. Specimen temperature, stiffness reduction
(normalized secant modulus) and electrical resistivity were simultaneously measured and correlated with fatigue life. During the test, the
maximum load in a fatigue cycle was increased in stepwise fashion.
Fig. 5 suggests the following:

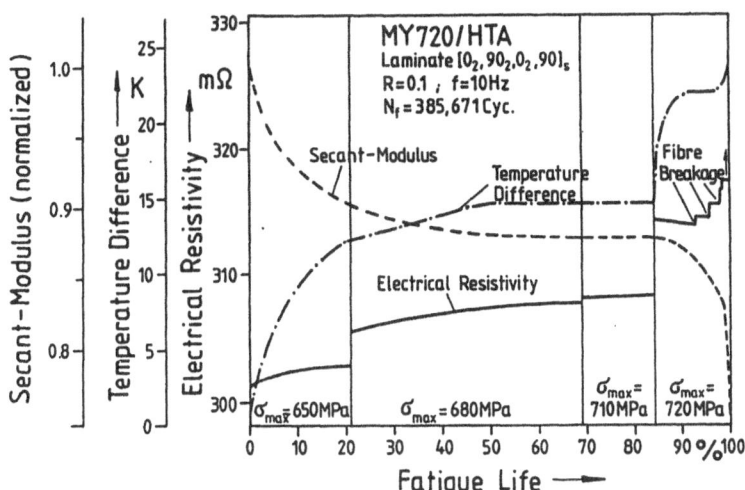

Fig. 5: Variation of secant modulus, temperature and electrical resistivity during cyclic loading (% of fatigue life) of a cross-ply laminate [1].

- During the test, a continuous increase in temperature, dependent on frequency and maximum load level, can be seen.
- The electrical resistivity also increases, but at each load step the resistivity jumps to a higher level.
- The high stiffness reduction at the beginning of the fatigue test can be related to transverse crack development in the 90°-plies, which is strongly related to the temperature increase.
- At the end of the fatigue test, approaching final failure, a strong stiffness reduction is again observed, accompanied by a rapid temperature increase. The resistivity jumps in steps to higher levels, which indicates that load bearing 0°-fibres have failed. The onset of final failure is then initiated.

It was shown in previous papers that the final reduction in stiffness can be related to the failure of a single fibre or even bundles of fibres [3].

The method of monitoring the electrical resistivity is thus an appropriate non-destructive in-situ technique for the investigation of fibre fracture.

Fig. 5 shows the temperature increase during the entire fatigue life.
The electrical resistivity increased with each step in the fatigue
load level. At the end of the fatigue life, during cycling at the
maximum fatigue load level of σ_{max} = 720 MPa, the resistivity of the
specimen decreases as a result of the temperature increase. Erratic
increases in resistivity occur only in association with fibre fracture. The observed decrease in the electrical resistivity can clearly
lead to misinterpretation of the results.

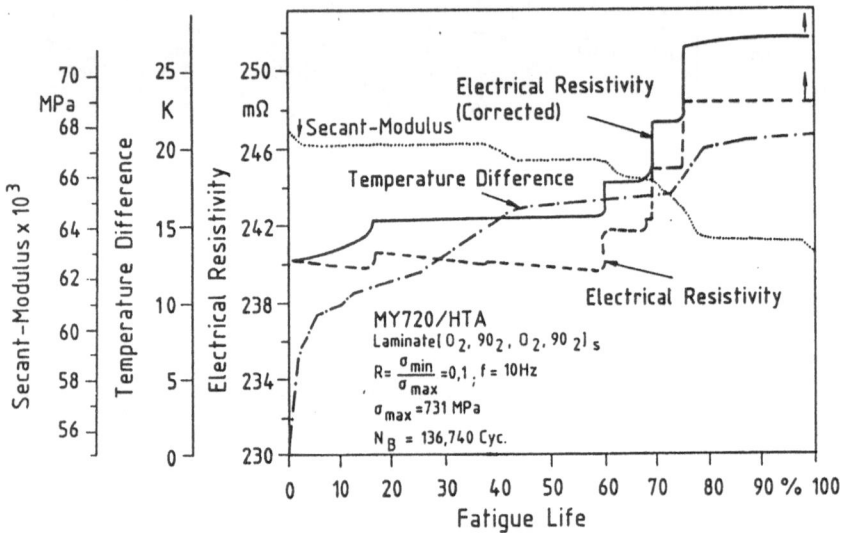

Fig. 6:   Variation of secant modulus, temperature and electrical
            resistivity during cyclic loading (% of fatigue life).
            Electical resistivity corrected for temperature variation
            [1].

A correlation to the resistivity on the basis of temperature measurements is therefore necessary. The variation of the resistivity due to
the temperature increase of a cross-ply test piece has to be identified. From these data the equivalent resistivity values can be taken
to correct the actual resistivity data of the fatigue test. The cor-
rected values for the electrical resistivity for another fatigue test
are shown in Fig. 6. A continuous increase, or at least a constancy
in the electrical resistivity can be observed, but there is no longe
any decrease.

4. Conclusion

It could be shown, that the measurement of the temperature variation can be used to get information about the internal damage situation in a carbon fibre reinforced plastic laminate. It can especially be used in fatigue tests as a damage analogue. The actual temperature of a test coupon has also to be detected, if the variation of the electrical resistivity in a carbon fibre reinforced plastic laminate is used as a damage analogue. A proper information can only be achieved by correcting the resistivity information with the temperature.

Literature

[1] K. Schulte, Ch. Baron: Load and failure analyses of CFRP-laminates by means of electrical resistivity measurements. Composite Science and Technology 36 (1989) 63-76

[2] K. Schulte, W.W. Stinchcomb: Damage mechanisms - including edge effects - in carbon fibre-reinforced composite materials. In: Application of Fracture Mechanics to Composite Materials, edited by K. Friedrich, Elsevier Science Publisher, Amsterdam (1989) 273-325

[3] K. Schulte: Damage development under cyclic loading. Proc. European Symposium on Damage Development and Failure Processes in Composite Materials, 4.-6. May 1987, Leuven, Belgium, (1987) 39-54

[4] K. Schulte: Stiffness reduction and development of longitudinal cracks during fatigue loading of composite laminates. Proc. European Mechanics Colloquium 182, "Mechanical Characterisation of Load Bearing Fibre Composite Laminates", 29.-31. Aug. 1984, Elsevier Science Publisher, London (1985) 36-54

[5] H. Neubert, H. Harig, K. Schulte: Monitoring of fatigue induced damage processes in cfrp by means of thermometric methods. Proc. ICCM-VI/ECCM-2, 20.-24. July, London. Elsevier Science Publisher (1987) 1.359-1.368

[6] H. Neubert, Ch. Baron: Thermometric measurements and electrical resistivity measurements for monitoring fatigue damage processes in cfrp. Proc. European Symposium on Damage Development and Failure Processes in Composite Materials, 4.-6. May 1987, Leuven, Belgium, 107-113

[7] H. Neubert: Anwendung thermometrischer Methoden bei der Prüfung kohlenstoffaserverstärkter Kunststoffe. Dr.-Ing. Dissertation, Universität-GH Essen, April 1989

[8] H. Neubert, K. Schulte, A. Harig: Evaluation of the fatigue behaviour by monitoring continuously temperature development in CFRP-laminates. ASTM-STP 1059 (1990) 435-453

[9] W. Thomson: The Quartely of Pure and Applied Mathematics 1 (1857) 57-77

[10] W. Thomson: Mathematical and Physical Papers 1 (1882), Collected Works, Cambridge University Press, Cambridge (1882) 174-332

A STATISTICAL APPROACH TO THERMAL SHOCK BEHAVIOUR OF SYNTACTIC FOAM

Dan Wei, D. Baptiste, Ph. Bompard and D. François
CNRS URA 850 Ecole Centrale de Paris
Grande Voie des Vignes F 92295 Châtenay-Malabry

Abstract. The failure characteristics of a syntactic foam made of hollow glass microspheres bounded by an epoxy resin matrix can be described by a Weibull statistical approach. A serie of tensile tests gave two Weibull parameters. It was checked that this approach allowed to predict the fracture strength of notched specimens as well as the fracture toughness K_{IC}. In the case of a thermal shock, it can be shown that the time at which failure occurs is not only a function of the thermal characteristics of the material, but depends also on the Weibull exponent.

1.Introduction.

Syntactic foam, a material made of hollow glass microspheres bounded by an expoxy resin, is particularly prone to thermal shock failure because it possesses a rather low thermal conductivity and a high coefficient of thermal expansion. This represents a problem in the fabrication of this material, as the polymerization generates heat and this results in thermal gradients, inducing tensile stresses at the surface. If the process is not well controleed failure of the part might take place.

We studied the tensile properties of this material, and we drew a micromechanical model to relate the macroscopic mechanical properties to the microstructure, namely to the distribution of the sizes of the microspheres (1, 2, 3). We used Weibull statistics to describe the tensile properties and we were thus able to predict the failure of notched specimens and the fracture toughness KIC of this material.

We will first recall these results, in order to show how such a model can be used in thermal shocks situations.

2. Material

The syntactic foam studied contained hollow glass spheres whose diameter ranged between 20 and 200 μm. with a wall thickness of 1 to 2 μm. The size followed a log normal distribution. Their total volume fraction was about 0.64.

The mechanical properties were the following :

density : 0.63

Young's modulus 2700 MPa. Poisson ratio 0.30

tensile strength 28 MPa

compressive strength 85 MPa

fracture toughness K_{IC}. 0.78 MPa $\sqrt{m}$

22 tensile specimens displayed a dispersion of the fracture stress which could be described with a two parameters Weibull plot

$$P_R = 1 - \exp\left[-\left(\sigma/\sigma_u\right)^m \frac{V}{V_0}\right]$$

In this expression σ is the fracture stress, V the volume of the specimen and P_R the fracture probability.

The Weibull parameters were found to be

m = 13.9

σ_u = 60.0 MPa

V_0 = 0.216 mm³

This value of V_0 was chosen to be the volume of a cube of side equal to 3 times the biggest diameter of the microspheres (or 15 times the mean diameter) (fig. 1).

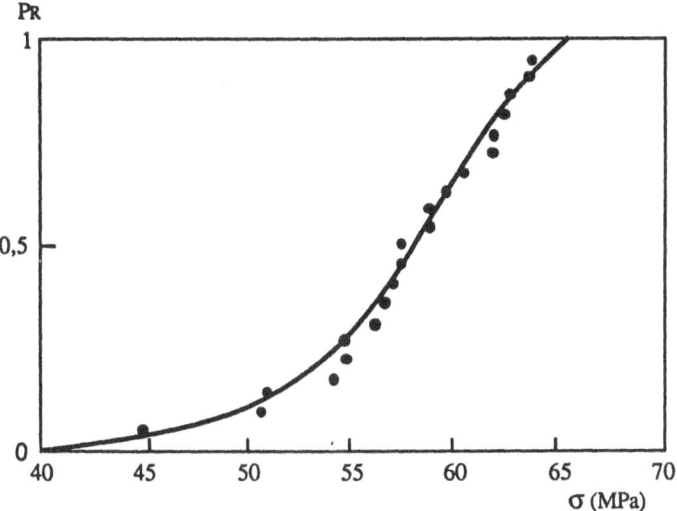

Figure 1.   Weibull distribution of the tensile fracture stress of the syntactic foam : fracture probability as a function of the fracture stress.

The size distribution function of the defects could be related to the Weibull parameters by the relation.

$$f(\phi) = \left(\pi/2 \; \alpha^2\right)^{m/2 + 1} \left(m/\pi V_0\right)\left(K_{Ic}/\sigma_u\right)^m /\phi^{m/2 + 1}$$

Where ϕ is the diameter of the microspheres and α an interaction parameter which was adjusted at a value of 1.4.

Figure 2 shows the fit which was obtained between the experimental size distribution and the one given by the above formula derived from the Weibull law.

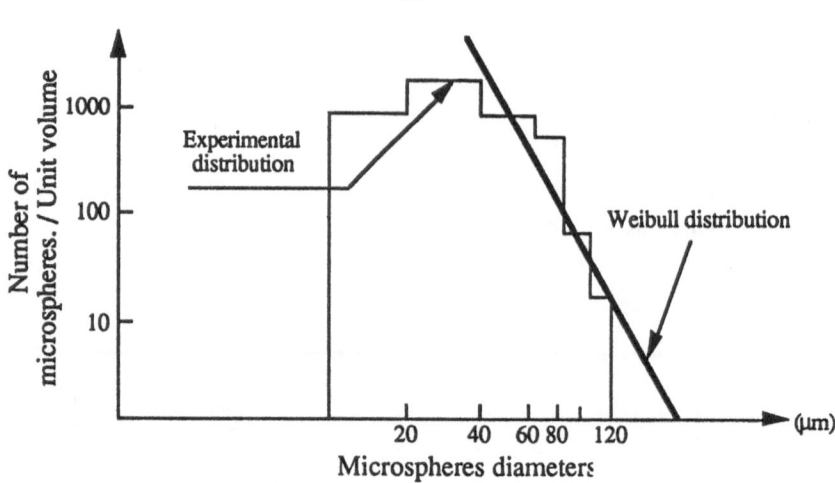

Figure 2.   Comparaison between the size distribution of the microspheres and the size of the defects deduced from the Weibull distribution of the fracture stress.

3. Use of the Weibull law to predict the failure probability of parts and relation with K_{IC}..

When the stress is not uniform the failure probability can be calculated with the formula

$$P_R = 1 - \exp\left[- \int (\sigma/\sigma_u)^m \frac{dV}{V_0} \right]$$

It is convenient to define the Weibull stress σ_w such that

$$\sigma_w^m = \int_v \frac{\sigma_I^m}{V_0} dV$$

Where σ_I is the maximum principal stress and the integral is taken over the volume where it is positive.
As an example we tested five notched specimens described on figure 3. The fracture load is given in table I and the corresponding Weibull stress was computed using a finite elements model. This allowed to show that these experimental results all fell in the scatter band of 10 to 90% predicted probability of failure.

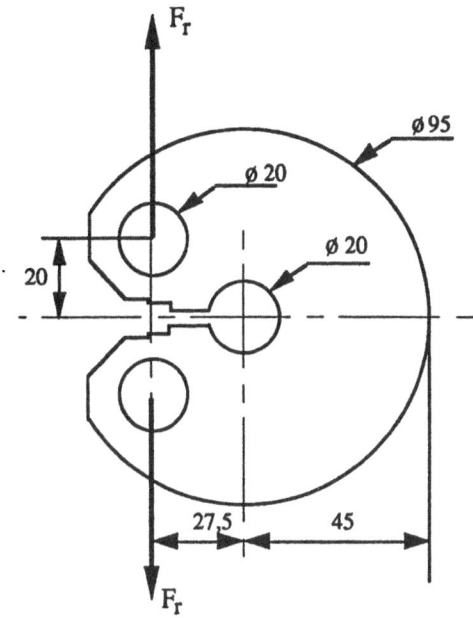

Figure 3.   Shape of the notched specimens.

Table I.   Fracture load of the notched specimens and corresponding Weibull stress and failure probability.

Test number	1	2	3	4	5
F_R (Kg)	900	890	870	856	725
σ_w (MPa)	63.5	62.8	61.4	60.4	51.2
P_f	0.89	0.86	0.75	0.67	0.12

Furthermore using the stress distribution in the elastic singularity of a crack, from the Weibull law of the material the fracture toughness could be deduced as :

$$K_{IC} = \left[\frac{4/\pi}{\left[1 + \frac{35}{8} \frac{m-3}{(m-2)(m-4)} \right] B \, r_{co}^2} \, V_0 \right]^{1/m} \sigma_u \sqrt{\pi \, r_{co}} \left[Ln \left(\frac{1}{1-P_R} \right) \right]^{1/m}$$

Where B is the crack front length

and $r_{co} = \left(\pi/3\sqrt{2} \, \eta_0 \right)^{1/3} \phi_0$

ϕ_0 being the largest diameter of the unbroken microspheres when the fracture strength of the resin is reached and η_0 the corresponding volume fraction, respectively 25 µm and O.2.

Table II. Shows a comparaison between this theoretical value and the experimental results.
Table II. Fracture toughness values and failure probabilities.

Pa	0.1	0.5				0.9
K_{IC} MPa $\sqrt{m}$ theory	0.69	0.79				0.85
K_{IC} MPa $\sqrt{m}$ experiment		0.70	0.78	0.79	0.83	

The measured fracture toughness values all fall within the 10% - 90% bounds of the failure probability and are close to the predicted value for a failure probability of O.5.

4. Application to thermal shock resistance.

Let us assume that the surface of a thick plate of syntactic foam is suddenly cooled from T_0 to T_I ($T_0 - T_I = \Delta T$).
The temperature distribution follows a variation which is an error function of the variable $x/2\sqrt{Kt}$, x being the de pth coordinate, t the time and $K = k/\rho C$ where k is the thermal conductivity, C the heat capacity and r ρ the specific mass. The thermal contraction of the material at the surface, assuming plane strain condition, induces tensile biaxial stresses which can start cracks on the surface. If a deterministic approach is used, this happens only if the temperature difference ΔT is high enough for the stress to exceed the fracture strength of the material and then cracking occurs at time $t = O$.
Now knowing the stress distribution from the evolution of the temperature, the Weibull stress σ_W can be computed, and hence the evolution of the failure probability, as a function of time. In general the integration needed to compute the Weibull stress cannot be achieved analyticaly. However a numerical method can give the result.
In order to obtain a feeling for the predicted trends, the problem can be simplifled by assuming a linear temperature distribution which penetrates in the plate as a function of 2 $\sqrt{Kt}$. Thus

$$T = T_0 + \Delta T \left(\frac{x}{2\sqrt{Kt}} - 1 \right) \quad \text{if } x \le 2\sqrt{Kt}$$

$$T = T_0 \quad \text{if } x > 2\sqrt{Kt}$$

The corresponding biaxial stress field is

$$\sigma = \frac{2E\,\alpha\,\Delta T}{(1+v)(1-2v)}\left[1 - \frac{x}{2\sqrt{Kt}}\right] \quad \text{if } x \le 2\sqrt{Kt}$$

$$\sigma = 0 \quad \text{if } x > 2\sqrt{Kt}$$

Where E is Young's modulus, V the Poisson ratio, α the coefficient of thermal expansion
The Weibull stress is then given by

$$\sigma_w^m = \frac{S}{V_0} \int_0^{2\sqrt{Kt}} \left[\frac{2E\,\alpha\,\Delta T}{(1+v)(1-2v)} \right]^m \left(1 - \frac{x}{2\sqrt{Kt}}\right)^m dx$$

for a plate of area S.

Finaly the failure probability is such that

$$Ln\frac{1}{1-P_R} = \left[\frac{2E\,\alpha\,\Delta T}{(1+v)(1-2v)} \right]^m \frac{S}{V_0\,\sigma_u^m} \frac{2\sqrt{Kt}}{m+1}$$

Assume $= 30.10^{-6}\ K^{-1}$ and $4K = 10^{-6}\ m^2\ sec^{-1}$.

If $T = 100°C$, it is found that the time for a crack to start in a plate 10 x 10 cm^2, is 4 sec with a failure probability of 50%, whereas the maximum stress at the surface reaches only 15.6 MPa, below the fracture strength.

The fracture probability increases as a function of time. Owing to the assumption of plane strain condition the entire plate will be under stress when the temperature becomes constant and equal to T_I. In more realistic calculations it will be found that the stress increases at first reaches a peak and then decreases as a function of time. Thus the failure probability will reach a maximum for a given time.

Given a certain probability of failure, say 50%, the relative variation of the fracture time is given by

$$\frac{\Delta t}{2t} = m\frac{\Delta\sigma_u}{\sigma_u} + \left\{ \frac{1}{m+1} - Ln\left[\frac{2E\,\alpha\,\Delta T}{(1+v)(1-2v)\,\sigma_u} \right] \right\} \Delta m$$

yielding

$$\frac{\Delta t}{t} \approx 28\frac{\Delta\sigma_u}{\sigma_u} + 20\frac{\Delta m}{m}$$

Thus the fracture time increases with both Weibull parameters, and a small gain can bring a large improvement. To increase the average fracture stress σ_u the average size distribution of the microspheres should be diminished because the stress concentration in the glass walls was found to vary as the square root of the microsphere diameter. To increase the Weibull exponent m the size distribution of the microspheres should be narrower. This will be limited by the scatter in the fracture properties of the glass itself which was not incorporated in this analysis, but which would add another Weibull distribution to the failure strength.

However this conclusion would be completely different for severe thermal shocks, with a temperature drop so large that the logarithm in the above expression, giving the relative variation of the time to fracture as a function of the relative variation of the Weibull exponent, would become larger than $1/(m+1)$. Adopting again the same values for the various parameters, this would correspond to a temperature drop larger than 200°C. In such a case, a widening of the size distribution of the microspheres would be beneficial provided the average would remain as low as possible.

Conclusion :

Considering that the tensile fracture strength of syntactic foam obeys a Weibull distribution which is related to the size distribution of the glass microspheres and that this statistical approach was able to predict the failure probability of notched or cracked parts, it was shown that a better resistance to weak thermal shocks could be achieved by increasing the Weibull exponent and the average fracture stress. This would be obtained by narrowing the size distribution of the microspheres keeping the average as small as possible. For strong thermal shocks on the contrary, the size distribution should be as wide as possible while the conclusion would remain the same as far as the average size is concerned.

I. Dan Wei "Etudes micromécaniques de l'endommagement des mousses syntactiques". These. Ecole Centrale des Arts et Manufactures (1987).

2. Dan Wei, D. Baptiste, Ph. Bompard, D. Francois. "Damage Micromechanisms Approach of an Heterogeneous Material". Proc. ECF 7. European Conference on Fracture. Budapest (1988).

3. Dan Wei, D. Baptiste, Ph. Bompard, D. François. "Statistical Failure Approach to an Heterogeneous Material" ICF 7. International Conference on Fracture. Houston (1989).

Static and Dynamic Pull out of an Elastic Rod from a Rigid Wall

Ingo Müller   W. Müller   P. Villaggio

Physikalische Ingenieurwissenschaft   Scienza di Costruzioni

TU Berlin, 1000 Berlin 12   Univ. di Pisa, 56100 Pisa

1. Quasistatic Pullout by Energy Criterion

The simplest imaginable fracture problem occurs in the pull-out of a linearly elastic rod that is glued into a rigid wall, see Figure 1. L_0 and $L = L_0 + s$ are the undistorted lengths of the undetached rod and of the partially detached one. l_0 and $l = l_0 + s$ are the corresponding lengths of the distorted rod. The distorsion is given by the elongation Δ of the free end such that $l_0 = L_0 + \Delta$ and $l = L + \Delta$ holds.

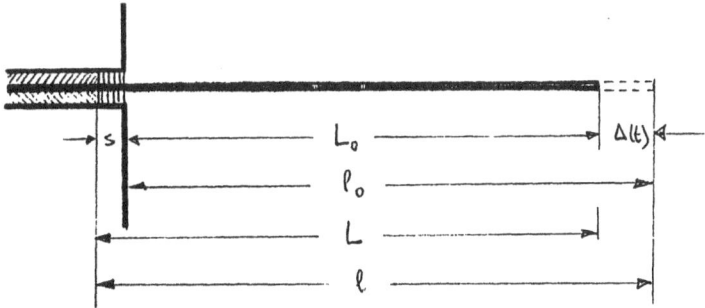

Figure 1: Rod glued into a rigid wall

We let X and x be the distance of a particle of the rod from the wall in the undistorted and distorted states respectively, so that $u(X,t) = x(X,t) - X$ is the displacement. As is well-known a displacement propagates along the rod obeying the wave equation of linear elasticity, viz.

$$\ddot{u} - c^2 \frac{\partial^2 u}{\partial X^2} = 0 \quad \text{with} \quad u(L_0,t) = \Delta(t) \quad \text{and} \quad u(-s,t) = 0. \tag{1.1}$$

$c = \sqrt{\frac{\lambda+2\mu}{\varrho}}$ is the speed of sound, where λ and μ are the Lamé coefficients of the rod and ϱ is its density.

In a quasi-static displacement the acceleration is neglected so that (1.1) reduces to the ordinary differential equation

$$\frac{\partial^2 u}{\partial X^2} = 0. \quad \text{Hence} \quad u(X,t) = \frac{\Delta(t)}{L_0 + s} (X + s). \tag{1.2}$$

The stress in the rod is given by

$$\sigma = E \frac{\partial u}{\partial X} \quad \text{with} \quad E = \frac{\mu(3\lambda + 2\mu)}{\lambda + \mu} \quad - \text{ elastic modulus}$$

and it is constant along the rod. The force needed for the elongation $\Delta(t)$ is therefore given by

$$F(\Delta, s) = \sigma\, A = \frac{EA}{L_0 + s}\, \Delta \tag{1.3}$$

where A is the cross-section of the rod.

We obtain the elastic energy stored in the rod by integration

$$W_{el}(\Delta, s) = \int_0^\Delta F(x, s)\, dx = \frac{EA}{2}\, \frac{\Delta^2}{L_0 + s} \;. \tag{1.4}$$

In the partially detached rod we also have to consider the energy stored in the detached area. This is akin to the surface energy of a newly opened crack and we take it to be proportional to s, the length of detachment. Therfore the total energy reads

$$W(\Delta, s) = \frac{EA}{2}\, \frac{\Delta^2}{L_0 + s} + \gamma\, s \;. \tag{1.5}$$

Figure 2 shows W as a function of s with Δ as parameter on the left and as a function of Δ with s as parameter on the right. Both groups of curves lend themselves for a determination of the critical elongation Δ_{cr}, where the detachment starts, and for the calculation of the length of detachment as a function of the elongation Δ.

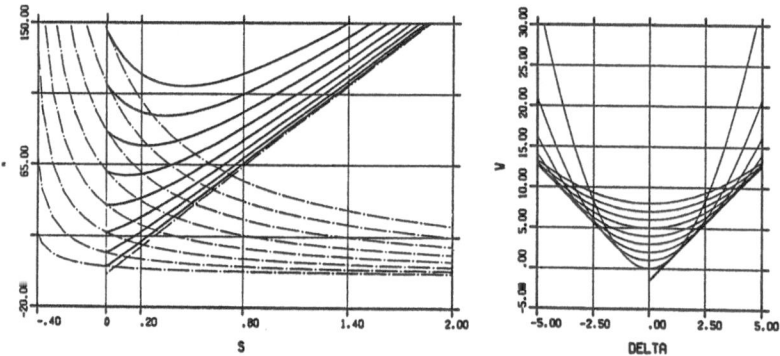

Figure 2: Energy as a function of Δ and s.

On the left hand side of Figure 2 we have plotted the energy as a superposition of the hyperbolae $\frac{EA}{2}\, \frac{\Delta^2}{L_0 + s}$ and of the straight line γs. The abscissa of the minimum of the superposition in the range $s \geq 0$ determines the length of the detachment. It is easily seen that for $\Delta \leq \Delta_{cr} = \sqrt{\frac{2\gamma}{EA}}\, L_0$ that minimum lies at $s = 0$ while for larger values of s the minimum lies at

$$s = \sqrt{\frac{EA}{2\gamma}}\, \Delta - L_0 \qquad \text{for} \qquad \Delta \geq \Delta_{cr} = \sqrt{\frac{2\gamma}{EA}}\, L_0 \;. \tag{1.6}$$

On the right hand side of Figure 2 the energies are plotted as parabolae with vertices on the ordinate at γs and with decreasing curvature $\dfrac{EA}{L_0 + s}$ as s increases. Therefore the parabolae intersect each other and, of course, the detachment will adjust itself so as to provide smallest energy, i.e. so as to be on the lowest parabola for any given Δ. The first intersection occurs at $\Delta_{cr} = \sqrt{\dfrac{2\gamma}{EA}}\ L_0$ and from there on the detachment will have the value s that coincides with the envelope of the group of parabolae. The envelope is calculated by elimination of s between

$$W = \frac{EA}{2}\ \frac{\Delta^2}{L_0 + s} + \gamma s \qquad \text{and} \qquad \frac{\partial W}{\partial s} = -\frac{EA}{2}\ \frac{\Delta^2}{(L_0 + s)^2} + \gamma = 0\ .$$

We conclude that the envelope is given by the straight line

$$W_{ENV} = \sqrt{2\gamma EA}\ \Delta - \gamma\ L_0\ . \tag{1.7}$$

For a given Δ we therefore find the corresponding value of s as the one whose energy (1.5) has the slope of the envelope, viz. $\sqrt{2\gamma EA}$. Once again we thus obtain (1.6).

The equation (1.6) determines the length s of detachment as a function of the elongation $\Delta(t)$, but only in the quasistatic approximation. If we wish to consider dynamic detachment, we need more sophistication in the description of rod and glue; it is then not enough to characterize the energy of detachment. Therefore we proceed to describe a structural model of the system.

2. A Model and its Critical Elongation in Quasi-static Displacements

The model shown in Figure 3 represents the glue by an array of grips and springs. The grips are a distance δ apart and the springs are elastic with spring constant λ. They are supposed to break, if they are elongated to the critical length β_{cr}. Obviously the springs shield the part of the rod behind the grips from the full strength of the stress in the main part of the rod. We shall assume that the shielding is so effective that no displacement occurs in the rod behind the second intact spring.[*]

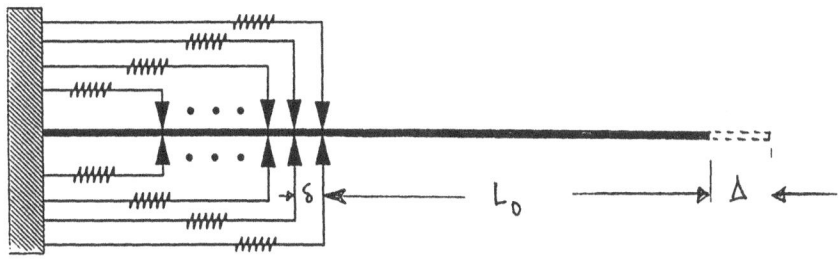

Figure 3: Modelling rod, glue, and rigid wall.

[*] This assumption will be relaxed in a future work. It simplifies the analysis greatly and therefore it is made here.

With the assumptions as described, only two parts of the rod have any displacement at all, viz. part 1 with $-\delta \leqslant X \leqslant 0$ and part 2 with $0 \leqslant X \leqslant L_0$. For the quasi-static case the displacements are linear functions of X, because they must satisfy the differential equation $(1.2)_1$. Before any spring has broken we have the boundary conditions

$$u_1(-\delta) = 0, \qquad\qquad u_1(0) = u_2(0),$$

$$EA \left.\frac{\partial u_1}{\partial X}\right|_{X=0} + \lambda u_1(0) = EA \left.\frac{\partial u_1}{\partial X}\right|_{X=0}, \qquad u_2(L_0) = \Delta. \tag{2.1}$$

Equation $(2.1)_3$ represents the balance of the elastic forces in the rod and in the spring at the first grip. The other equations are self-evident. With (2.1) the displacements are easily calculated to be

$$u_1 = \beta\left(\frac{X}{\delta} + 1\right)$$

with $\qquad \beta = \dfrac{\Delta}{1 + \dfrac{L_0}{\delta}\left(1 + \dfrac{\lambda\delta}{EA}\right)}.$ $\tag{2.2}$

$$u_2 = -\beta\left(\frac{X}{L_0} - 1\right) + \frac{\Delta}{L_0} X$$

β is the elongation of the spring and $(2.2)_3$ shows how that is related to the elongation of the rod. The left hand side of Figure 4 shows u(X). The kink in the curve represents the shielding of the rod behind the spring.

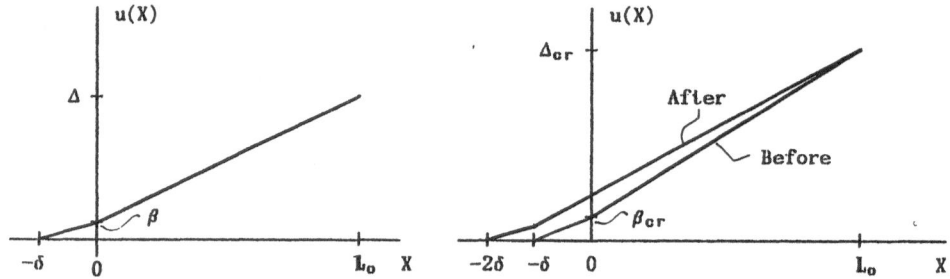

Figure 4: Displacement in the rod before and after first break.

Of course, a glue grips the rod continuously while the grips of the model are the distance δ apart according to Figure 3. Therefore we have to assume that $\delta \ll L_0$ holds, if the model is to be realistic. In that case it follows from $(2.2)_3$ that

$$\frac{\beta}{\delta} \xrightarrow[\frac{\delta}{L_0} \ll 1]{} \frac{\frac{\Delta}{L_0}}{1 + \frac{\delta\lambda}{EA}}.$$

For shielding we need to have $\frac{\beta}{\delta} < \frac{\Delta}{L_0}$ and therefore we must require that $\delta\lambda$ stays finite as δ tends to zero.

The right hand side of Figure 4 shows the displacements of the rod and the spring for $\Delta = \Delta_{cr}$ before and after the break of the spring has occurred. The energies of the spring and of the two parts of the rod are easily calculated, the latter ones from (1.4), and they are represented in the following table.

BEFORE		AFTER	
$W_S = \frac{\lambda}{2}\,\beta_{cr}^2$	$= \frac{\lambda}{2}\,\dfrac{\Delta_{cr}^2}{\left[1+\frac{L_0}{\delta}\left(1+\frac{\lambda\delta}{EA}\right)\right]^2}$	$W_S = \frac{\lambda}{2}\,u^2(-\delta)$	$= \frac{\lambda}{2}\,\dfrac{\Delta_{cr}^2}{\left\{2+\frac{\lambda\delta}{EA}+\frac{L_0}{\delta}\left(1+\frac{\lambda\delta}{EA}\right)\right\}^2}$
$W_1 = \frac{EA}{2}\,\dfrac{\beta_{cr}^2}{\delta}$	$= \frac{EA}{2\delta}\,\dfrac{\Delta_{cr}^2}{[\quad]^2}$	$W_1 = \frac{EA}{2}\,\dfrac{u^2(-\delta)}{\delta}$	$= \frac{EA}{2}\,\dfrac{\Delta_{cr}^2}{(\quad)^2}$
$W_2 = \frac{EA}{2}\,\dfrac{(\Delta_{cr}\mp\beta_{cr})^2}{L_0} = \frac{EA}{2L_0}\,\Delta_{cr}^2\left(1-\frac{1}{[\;]}\right)^2$		$W_2 = \frac{EA}{2}\,\dfrac{(\Delta_{cr}\mp u(-\delta))^2}{L_0+\delta} = \frac{EA}{2(L_0+\delta)}\,\Delta_{cr}^2\left(1-\frac{1}{\{\;\}}\right)^2$	

From the table we are able to determine $W = W_S + W_1 + W_2$ before and after the break and calculate $W^B - W^A$ as a function of Δ_{cr} or vice-versa. In the case $\delta \ll L_0$ we obtain

$$\Delta_{cr} = \sqrt{\frac{2(W^B - W^A)/\delta}{EA}}\; L_0 \; . \tag{2.3}$$

Comparison with $(1.6)_2$ shows that the model provides the same formula for Δ_{cr} as the argument presented in section 1, which is the standard argument of fracture mechanics involving surface energy. The only difference between $(1.6)_2$ and (2.3) is that the surface energy per crack length γ is now replaced by the energy release $(W^B - W^A)/\delta$ per "crack length" δ. Of course, most of the energy $W^B - W^A$ will go into heat as the rod jumps from one displacement to the other one in Figure 4. But this is the same with γs in section 1.

By the above quasi-static arguments we have gained some confidence into the model and now we proceed to a proper dynamic situation.

3. Dynamic Pullout

Now we take the full wave equation (1.1) seriously and assume that the elongation $\Delta(t)$ of the free end of the rod creates waves when displacements strain the springs and possibly make them break. The general solution of the wave equation (1.1) reads

$$u(X,t) = \theta_1(X - c(t - \tau)) + \theta_2(X + c(t - \tau)) \tag{3.1}$$

where θ_1 and θ_2 are calculated from the initial displacement $u(X,\tau)$ and rate of displacement $\dot{u}(X,\tau)$ as follows

$$\theta_{\frac{1}{2}}(\alpha) = \frac{1}{2}\,u(\alpha,\tau) \mp \frac{1}{2}\frac{1}{c}\int \dot{u}(\alpha,\tau)\,d\alpha \; . \tag{3.2}$$

Thus the solution (3.1) is a superposition of two waves running with speed c to the right and left.

When i springs have been broken, the last one at time τ_i we have to find $u(X,t)$ from the displacement $u(X,\tau_i)$ and the rate of displacement $\dot{u}(X,\tau_i)$ at the time τ_i. As the wave develops, it must satisfy at all times the boundary conditions

$$u_1(-(i+1)\delta) = 0 \ ,$$

$$u_1(-i\delta) = u_2(-i\delta) \ ,$$

$$EA \left.\frac{\partial u_1}{\partial X}\right|_{X=-i\delta} + \lambda\, u_1(-i\delta) = EA \left.\frac{\partial u_2}{\partial X}\right|_{X=-i\delta} \ , \tag{3.3}$$

$$u_2(L_0) = \Delta(t) \ .$$

Here u_1 and u_2 as before denote the displacement on the left and right of the loaded spring at $X = -i\delta$.

A moment's reflection shows that the propagation of the wave to right and left, the transmission and reflection at $X = -i\delta$ and $X = -(i+1)\delta$, the retransmission and re-reflection at those points presents an enormous "book-keeping problem" that can only be solved numerically. We have started to look at a few cases and they are presented in the figures 5 and 6.

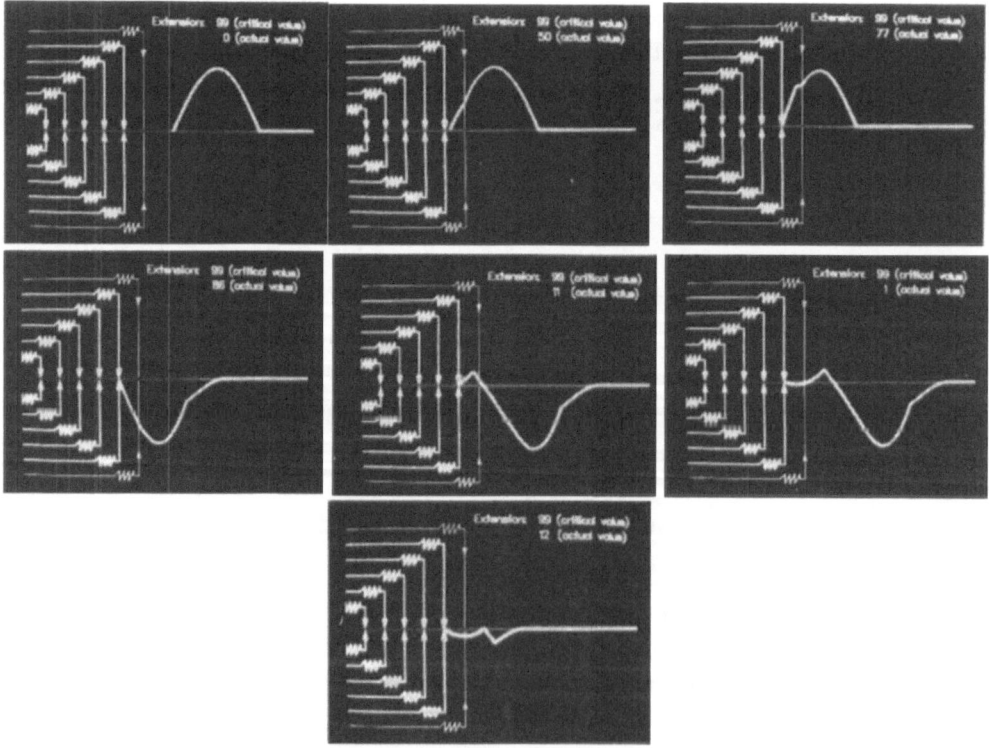

Figure 5: An incoming wave does not break springs

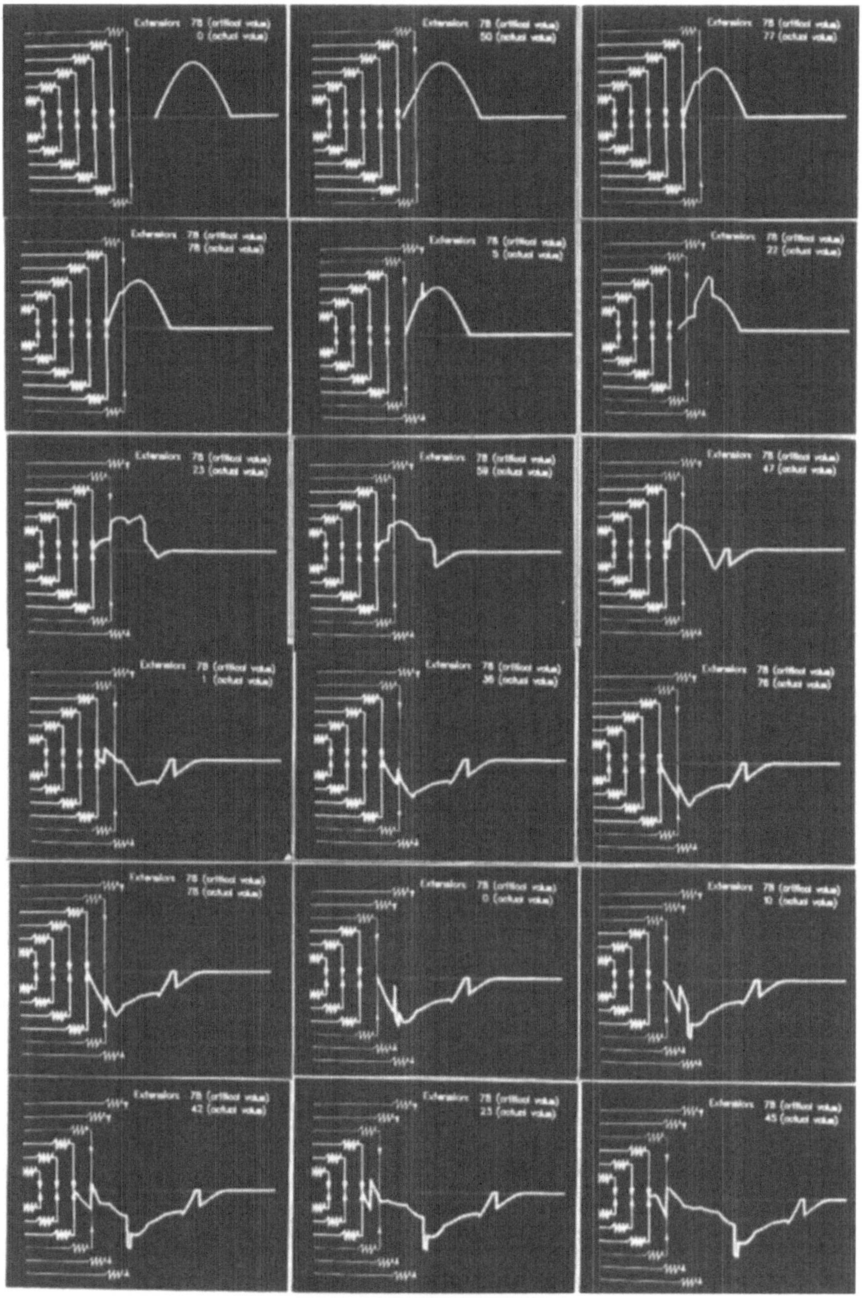

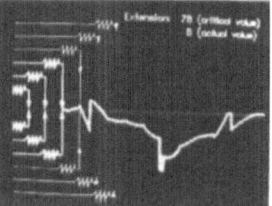

Figure 6: An incoming wave
breaks two springs before
moving back spent and dissipated.

In both figures we start with an incoming sine-shaped curve of height 120 in suitable non-dimensional units. The critical displacement β_{cr} is only 99 in the same units in Figure 5. Yet, no spring breaks as we can see. The reason is that the waves reflected at $X = 0$ and $X = -\delta$ weaken the incoming wave so much that the critical load is never reached on the first spring.

Figure 6 corresponds to a "weaker glue", i.e. a set of springs whose critical displacements are $\beta_{cr} = 79$. Accordingly two springs are broken, one already by the incoming wave, the other one later on. At the end the wave moves off to the right, having spent its energy in breaking two springs and having dissipated energy by the various reflections and transmissions.

The pictures of Figures 5 and 6 represent our first calculations and are, if anything, of heuristic value only. The model awaits a systematic numerical study with realistic values of the data. We trust, however, that studies like these may furnish valuable information on detachment of fibers and whiskers in reinforced materials under impulsive loading.

AN EXTERNAL RADIAL CRACK IN A UNIT CELL
OF A FIBRE-REINFORCED COMPOSITE

K. Herrmann*, I. Mihovsky**, M. Usunova**

* University of Paderborn, Laboratorium für Technische Mechanik, Pohlweg 47-49, 4790 Paderborn, FRG
** Sofia University, Dept. Math. and Informatics, A. Ivanov Str. 5, 1126 Sofia, Bulgaria

ABSTRACT

The problem of an external radial crack in a thermally stressed unit cell of a fibrous composite with an elastic-perfectly plastic matrix material is considered. By using a modified Dugdale type crack model together with the weight function technique different influences like the mechanical properties of the phases, the cell geometry, the plastic zone developing around the fibre, and the crack length onto the plastic zone length at the crack tip are considered.

INTRODUCTION

The variety of composite materials and existing crack configurations in those structures gives rise to different specific interaction problems. The present paper deals with such a problem for a unidirectionally reinforced fibrous composite. The fibres are continuous, elastic, and perfectly bonded to an elastic perfectly-plastic matrix. The fibre volume fraction is relatively small. The problem concerns a composite unit cell consisting of a circular cylindrical fibre with a coaxial matrix coating. A straight longitudinal crack with a crack front parallel to the fibre axis enters the matrix along the radial direction from its outer surface. The crack length is relatively small compared with the cell radius. The loading is defined as resulting from a process of progressive matrix cooling. The elastic-plastic properties of the matrix material are considered as temperature independent. Under these conditions, an axisymmetrical stress state will arise in the uncracked unit cell. Further, the crack itself is of the opening mode type.

Some of the aspects involved in the problem have been already treated in a series of articles by Herrmann /1/ and Herrmann and Mihovsky /2-7/. In the present work the approach from the references /2-7/ has been extended to the problem of an externally (edge) cracked unit cell. The influence of the fibre volume fraction, constituents properties, load intensity, and crack length, repectively, on the development of the plastic zones surrounding the fibre as well as the crack tip has been analyzed and illustrated by numerical results.

1. CELL AND CRACK GEOMETRY

The cell is referred to cylindrical coordinates $\{r,\theta,z\}$ and is thereby axialsymmetric with respect to the z-axis. The fibre and matrix cross-sections occupy the circle $0 \leq r \leq r_f$ and the annulus $r_f \leq r \leq r_m$, respectively. For small crack lengths the plastic zone surrounding the fibre may be considered to occupy the annulus $r_f \leq r \leq R_c$ (as in the case of an uncracked unit cell). The crack is of the Dugdale type. Its length is l and the length of the thin plastic zone is s. Fig. 1 shows the cross-section of the cracked unit cell.

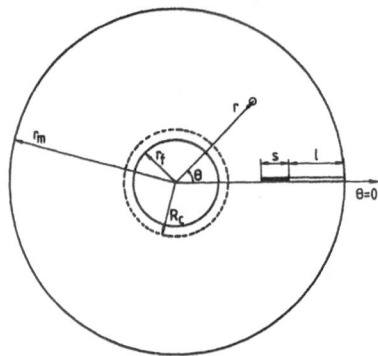

Fig. 1: Cross-section of the cracked unit cell.

Furthermore, the notations E_f, E_m and ν_f, ν_m are used for the Young's moduli and the Poisson's ratios of the fibre and the matrix material, respectively. The matrix material obeys the von Mises' yield condition and the associated flow rule (cf. for example /8/). Its tensile yield stress is σ_y. The E_m/E_f-ratio is relatively small compared to unity.

2. RELATED ARGUMENTS AND RESULTS

2.1 A GENERAL PLANE PROBLEM

A specific feature of the approach developed in the references /2-5/ is that it accounts for the interactions between the whole group of basic fibre reinforcement effects (stiffening, strengthening, shrinkage, and stress concentration) and the principal effect of the matrix ductility - its limited elastic response. In particular, the approach adopts that the elastic part of the axial strain is not negligible (due to the strong stiffening effect) and keeps an approximately constant average value ε^*. In the thermal case this adoption concerns not the total strain but only that part of the latter which is due to the thermal stresses. By means of this adoption the approach reduces the two model problems from the references /2,3/ to a general plane stress-like perfect plasticity problem /9/ with a yield condition of the form (ellipse in the plane of the principal cross-section stresses σ_r^{pl}, σ_θ^{pl})

$$(\sigma_r^{pl} - \sigma_\theta^{pl})^2 + (\sigma_r^{pl} + \sigma_\theta^{pl} - \frac{2E_m\varepsilon^*}{\sqrt{3}\tan\phi})^2 \tan^2\phi - \frac{4\sigma_y^2}{3} = 0 \tag{1}$$

where $\tan\phi = (1-2\nu_m)/\sqrt{3}$. The quantity ε^* is a measure of the above mentioned interactions and is specific for the given composite structure and the loading status as well.

The stresses within the plastified matrix region are

$$\left.\begin{matrix}\sigma_r^{pl}\\[6pt]\sigma_\theta^{pl}\end{matrix}\right\} = \frac{E_m\varepsilon^*}{1-2\nu_m} + \frac{\sigma_y}{\sqrt{3}\sin\phi}\cos(\omega\pm\phi) \tag{2}$$

where $\sin\omega = (\sigma_\theta^{pl} - \sigma_r^{pl})\sqrt{3}/2\sigma_y$. Thereby the angle $\omega = \omega(r)$ defines the positions of the stress states given by eqs. (2) along the yield ellipse. The cases $|\omega| < \phi$ (or $|\omega-\pi| < \phi$), $\phi < \omega < \pi - \phi$ (or $\phi < \omega - \pi < \pi - \phi$), and $|\omega| = \phi$ (or $|\omega - \pi| = \phi$) correspond (cf. references /8,9/) to elliptic, hyperbolic, and parabolic types of the set of equations governing the general plane problem.

2.2 MATRIX YIELDING IN CASE OF THERMAL LOADING

In the uncracked thermally loaded unit cell matrix yielding takes initially place at the interface $r=r_f$ due to the stress concentration effect. The angle $\omega_{r_f} = \omega(r_f)$ at this instant has the value ω_{R_C} where

$$\cos\omega_{R_C} = -\frac{E_m\varepsilon^*}{\sigma_y(1+\nu_m)} \tag{3}$$

Moreover, progressive matrix cooling implies a monotonic increase in ω_{r_f} due to increasing shrinkage, cf. eqs. (2) for $\sigma_r^{pl}(r_f)$ within the hyperbolic interval $[\omega_{R_C}, \pi - \phi]$. At the instant $\omega_{r_f} = \pi - \phi$ designating a parabolic point of the yield ellipse a free plastic flow or so-called plastic instability tends to take place at the interface. At this critical instant failure modes may also develop in the previous uncracked thermally loaded unit cell (debonding, matrix cracking, fibre breaking).

Further, the $\omega(r)$-dependence in the plastic zone is implicitly given by the relation

$$\frac{R_C^2}{r^2} = \frac{\sin\omega}{\sin\omega_{R_C}}\exp[(\omega - \omega_{R_C})/\tan\phi] \tag{4}$$

The plastic zone radius R_c increases in accordance with eq. (4), that means the plastic zone spreads into the matrix phase with progressive loading. Thereby, the radius R_c achieves its largest value $R_c^* < r_m$ which is given by eq. (4) with $r = r_f$ and $\omega = \pi - \phi$.

2.3 STRESS DISTRIBUTION

In the purely elastic state of the unit cell the transverse stresses in the matrix phase are

$$
\left.\begin{array}{c} \sigma_r^{el} \\ \sigma_\theta^{el} \end{array}\right\} = \frac{E_m}{1+\nu_m} \frac{C}{r_m^2} (1 \mp \frac{r_m^2}{r^2}) \tag{5}
$$

where

$$
C = -r_f^2 \alpha_m T_m (1+\nu_m) [1 - \frac{\nu_m - \nu_f}{(1+\nu_m)(1+E_c)}] \tag{6}
$$

$$
E_c = E_f r_f^2 / E_m r_m^2 \tag{7}
$$

$T_m \leqq 0$ is the uniform matrix temperature , α_m is the coefficient of linear thermal expansion of the matrix material.

Eqs. (5) apply to the stresses in the elastically deforming annulus $R_c \leqq r \leqq r_m$ as well, but with a new constant C^{ep}, which is

$$
C^{ep} = - \frac{1+\nu_m}{E_m} \frac{R_c^2}{1 - \frac{R_c^2}{r_m^2}} \sigma_r^{pl}(R_c) \tag{8}
$$

where $\sigma_r^{pl}(R_c)$ is given by eqs. (2) with $\omega = \omega_{R_c}$.

Matrix yielding takes place at a temperature $T_m = T_m^{pl}$ where

$$
T_m^{pl} = - \epsilon^* (1+E_c)/\alpha_m E_c \tag{9}
$$

The stresses in the plastic zone $r_f \leqq r \leqq R_c$ are given by eqs. (2) along with eq. (4) for the $\omega(r)$-dependence. The plastic zone radius R_c affects the stresses in both the elastic and the plastic regions through eqs. (5) (with C^{ep} instead of C) and (2) (with eq. (4) for $\omega = \omega(r)$), respectively. Its temperature dependence follows from the $R_c(\varepsilon_z)$ - and $\varepsilon_z(T_m)$ - relations which read according to the references /4,5/.

$$R_c^2 = R_c^{*2} [1 - (1 - \frac{r_f^2}{R_c^{*2}}) (1 - \frac{\Delta\varepsilon_z}{\Delta\varepsilon_z^*})] \tag{10}$$

$$\Delta\varepsilon_z = \alpha_m \Delta T_m [1 + E_c + \alpha_m \Delta T_m \frac{E_c}{1+E_c} \frac{R_c^{*2}}{r_m^2} (1 - \frac{r_f^2}{R_c^{*2}}) \frac{1}{\Delta\varepsilon_z^*}]^{-1} \tag{11}$$

In eqs. (10), (11) the notations $\Delta T_m = T_m - T_m^{pl}$, $\Delta\varepsilon_z = \varepsilon_z + \varepsilon^*/E_c$ are used, where ε_z is the total axial strain. The quantity $\Delta\varepsilon_z^*$ is the value of the $\Delta\varepsilon_z$-strain at the critical instant $T_m = T_m^*$ (with $R_c = R_c^*$, $\omega_{r_f} = \pi - \phi$).

2.4 THE MODIFIED DUGDALE CRACK MODEL

In the references /5,7/ it has been shown that the stress state within the thin Dugdale-type plastic zone at the tip of a radial crack in the matrix corresponds to the parabolic point $\omega = \phi$ of the yield ellipse, eq. (1). Using this value of ω eqs. (2) imply immediately that the stress σ_{DZ} acting across the plastic zone $(r_m - l - s) \leq r \leq (r_m - l)$ is (cf. also Fig. 1)

$$\sigma_{DZ} = \sigma_\theta \Big|_{\theta=0} = \sigma_\theta^{pl} \Big|_{\omega=\phi} = \frac{E_m \varepsilon^*}{1 - 2\nu_m} + \frac{\sigma_y}{\sqrt{3}\sin\phi} \tag{12}$$

3. THE EXTERNAL RADIAL CRACK PROBLEM

3.1 MODEL PROBLEM

In accordance with the assumptions about the cracked composite unit cell configuration (small fibre volume fraction and crack length) and following the very sense of Dugdale's approach one should expect that with an implicit reference to St. Venaint's principle a reliable approximate estimate of the crack tip plastic zone length might be derived from the solution of the following model problem.

The crack is considered as contained in a solid homogeneous circular cylinder of the matrix material. The stress state in the latter results from the σ_θ^{el}-stress distribution from eqs. (5) applied to the crack surfaces (with $C = C^{ep}$ in case of the presence of a plastic zone $r_f \leq r \leq R_c$). Such a model is, of course, valid if the condition $(l+s)/(r_m - R_c) < 1$, cf. Fig. 1, is satisfied, where l+s is the length of the fictitious crack appearing in the Dugdale approach. It is also clear enough that the cracked cylinder model becomes less realistic with progressive loading (matrix cooling), i.e. with increasing R_c- and s-values. But it should be noted at the same time that failure modes may develop in the unit cell at the critical instant $R_c = R_c^*$ (cf. Section 2.2), or due to a crack growth initiation at a critical plastic zone length $s^* = s^*(l)$ for which the ratios $(l+s)/(r_m-R_c^*)$ or $(l+s^*)/(r_m-R_c)$, respectively, are still small compared to unity. Thus, to consider the entire interval $0 < (l+s)/(r_m-R_c) \leq 1$ appears to be of interest although increasing $(l+s)/(r_m-R_c)$-ratios will definitely lead to a decrease of accuracy.

3.2 BOUNDARY VALUE PROBLEM FOR THE CRACKED CYLINDER MODEL

The cylinder has been considered under general plane strain conditions, i.e. $\varepsilon_z = \varepsilon^* + \alpha_m T_m$ is independent of r and z, respectively. Its lateral surface $r = r_m$ is stress-free. By means of the superposition principle and in accordance with the basic idea of Dugdale's approach the real edge crack of length l with stress-free surfaces is substituted by a fictitious crack of length $l+s$ with the following opening mode stress distribution applied to its surfaces (cf. Fig. 1):

$$\sigma_\theta(r,\theta)\Big|_{\theta=\pm o} = \begin{cases} \sigma_\theta^{el} & , \quad (r_m-1) \leqq r \leqq r_m \\ \sigma_\theta^{el} - \sigma_{DZ} & , \quad (r_m-1-s) \leqq r \leqq (r_m-1) \end{cases} \tag{13}$$

Thereby, the stress σ_θ^{el} in eqs. (13) is defined by eqs. (5) with C (cf. eq. (6)) or C^{ep} (cf. eq. (8)) in the cases of absence and presence of a plastic zone $r_f \leqq r \leqq R_c$, respectively. The stress σ_{DZ} is defined by eq. (12).

3.3 THE CRACK TIP PLASTIC ZONE LENGTH

Basing upon different arguments Bueckner /10/ and Rice /11/ have developed an effective approach, the so-called weight function method, to solve plane opening mode crack problems involving given cracked body geometry and arbitrary load systems. Further, the arguments and the results considered in the foregoing sections allow a determination of the plastic zone length s from the solution of the equation

$$K_I(s) = 0 \tag{14}$$

where $K_I(s)$ is the s-dependent stress intensity factor induced by the load system from eqs. (13). Thus, by applying the weight function method, eq. (14) takes the form

$$- \int_{r_m-1-s}^{r_m-1} \sigma_{DZ} \frac{\partial u_y}{\partial l} \, dr + \int_{r_m-1-s}^{r_m} \sigma_\theta(r,\theta)\Big|_{\theta=o} \frac{\partial u_y}{\partial l} \, dr = 0 \tag{15}$$

where u_y is the transverse displacement of the surfaces of the fictitious crack induced by a certain symmetrical load system.

Now an explicit form of eq. (15) can be obtained by using some approximate results presented by Gregory /12,13/ for the problem of a disc containing a radial edge crack subjected to a constant pressure /12/ or a concentrated force /13/, respectively.

A series of lengthy transforms and calculations based upon these results has implied the following explicit form of eq. (15):

$$
-\frac{\sigma_{DZ}}{2} S\left[\frac{(L+S)^{1/2}(4-2L-S)}{M(2-L-S)^{3/2}} + \frac{2(2-2L-S)}{N(L+S)^{1/2}(2-L-S)^{1/2}}\right] +
$$

$$
\frac{2A}{M}\frac{(L+S)^{1/2}}{(2-L-S)^{3/2}}\left[1+\frac{(2-L-S)^2}{4}\left(\frac{2M}{N}-1\right)\right] + A\left[\frac{(L+S)^{1/2}}{M(2-L-S)^{3/2}}\left(\frac{L+S}{1-L-S}-\ln(1-L-S)\right)-\right.
$$

$$
\left.\frac{2}{N(L+S)^{1/2}(2-L-S)^{1/2}}\ln(1-L-S)\right] = 0 \tag{16}
$$

where $L=l/r_m$, $S=s/r_m$, $M=0{,}355715...$, $N=0{,}966528...$, $A=E_m\tilde{C}/(1+\nu_m)r_m^2$ with $\tilde{C}=C$ and $\tilde{C}=C^{ep}$ for the linear-elastic and the elastic-plastic deformation of the unit cell, respectively.

It is noteworthy that in the particular cases when the stress distribution in eqs. (13) takes the form of a constant pressure or of a concentrated force, respectively, the left-hand side of eq. (16) implies the results obtained in /12,13/. In the case $l/r_m \to 0$ its left-hand side leads to the known result for a half-space with an opening mode edge crack. These implications support to a certain extent the validity of the latter equation.

4. NUMERICAL RESULTS

Eq. (16) has been solved for different cracked cell geometries (r_f/r_m- and l/r_m-ratios) and mechanical properties of the constituents (E_m/E_f-ratios and $\nu_m-\nu_f=\Delta\nu$-differences). The properties have not been referred to certain specific composite constituents. Nevertheless, all sets of properties considered involve the combination $E_m=6\cdot10^4 N/mm^2$, $E_f=24\cdot10^4 N/mm^2$, $\nu_m=0.31$, $\nu_f=0.27$, $\sigma_y=135 N/mm^2$, $\alpha_m=18.2\cdot10^{-6}K^{-1}$ which corresponds practically to a C-fibre/Al-matrix composite (with $T_f=0°C$, $T_m=-80°C$) and is involved in the computations carried out by Herrmann in /1/. All numerical results are obtained with the values of E_f, ν_f, σ_y and α_m specified above.

Thereby the solid circles (signs "•") in the figures 2-5 correspond to the instants of initial matrix yielding at the interface $r = r_f$. The star signs "*" reflect the instants at which the two plastic zones join each other.

For the sake of simplicity eq. (16) has been solved in the following way. For given s_i-lengths (starting from $s/l \approx 10^{-3}$) corresponding temperatures $T_{m,i}$ have been obtained from eq. (16). Then the $R_{c,i}(T_{m,i})$-values have been derived from the eqs. (10) and (11), respectively.

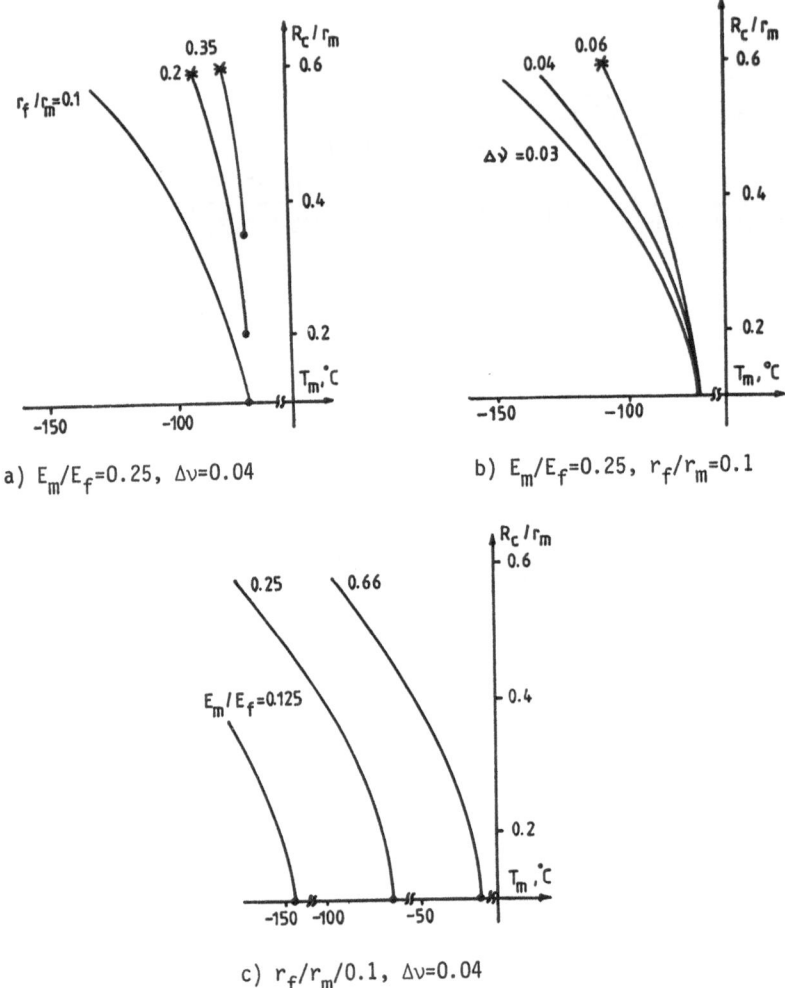

Fig. 2: The $R_c(T_m)$-curve for different parameters
a) fibre volume fraction; b) Poisson's ratios difference;
c) Young's moduli ratio

Fig. 2 shows the influence of a) the fibre-volume fraction r_f/r_m, b) the Poisson's ratios difference $\Delta\nu = \nu_m - \nu_f$, and c) the Young's moduli ratio E_m/E_f on the $R_c(T_m)$-curve. Fig. 2c reflects implicitly the σ_y-influence through the E_m/σ_y-ratios as well. The cases a) and b) prove that the increase in each of the r_f/r_m-and $\Delta\nu$-values implies increasing temperature sensitivity of the plastic zone radius. The T_m^{pl}-temperature is practically not influenced by the set of r_f/r_m- and $\Delta\nu$-values considered in these cases. Increasing E_m/E_f-ratios (or implicitly E_m/σ_y-ratios) lead to a translation of the $R_c(T_m)$-curve along the positive direction of the T_m-axis with a corresponding increase of the (negative) T_m^{pl}-temperature.

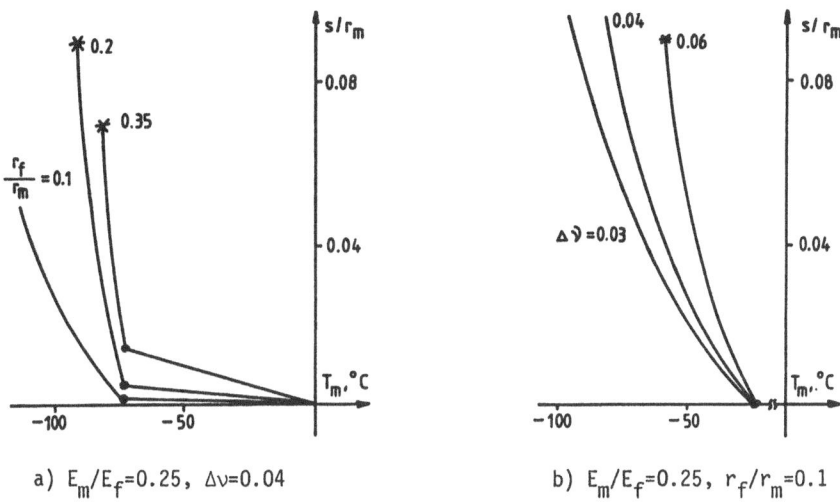

a) $E_m/E_f=0.25$, $\Delta\nu=0.04$ b) $E_m/E_f=0.25$, $r_f/r_m=0.1$

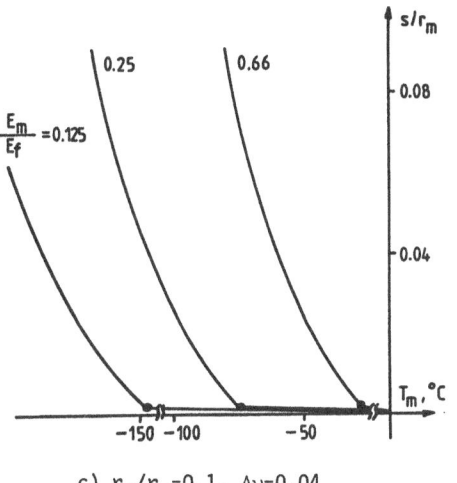

c) $r_f/r_m=0.1$, $\Delta\nu=0.04$

Fig. 3: The $s(T_m)$-dependence for different parameters
a) fibre volume fraction ; b) Poisson's ratios difference ;
c) Young's moduli ratio (in all cases valids l/r_m = 0.3)

Fig. 3 proves that the influence of the same set of parameters on the $s(T_m)$-curve is similar to that for the $R_c(T_m)$-curve. But in contrast to the latter case the $s(T_m)$-curves are concave. In other words the plastic zone length s is more sensitive to the temperature changes than the R_c-radius.

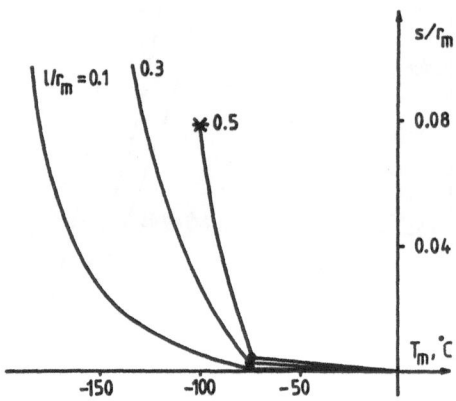

Fig. 4: The $s(T_m)$-curve for different crack lengths l
$(E_m/E_f = 0.25, \quad \Delta\nu = 0.04, \quad r_f/r_m = 0.1)$

The explicit $s(T_m)$-dependence for different crack lengths l is depicted in Fig. 4. The graphs are comparable with those from Fig. 3a.

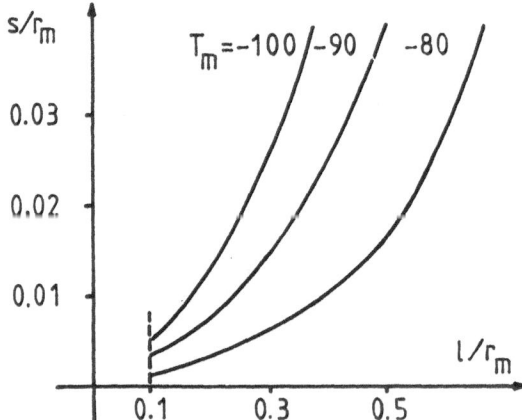

Fig. 5: The s(l)-curve at fixed values of T_m [°C]
$(E_m/E_f = 0.25, \quad \Delta\nu = 0.04, \quad r_f/r_m = 0.1)$

Further, Fig. 5 illustrates the s(l)-dependence at fixed T_m-values.

From the results shown in the figures 2 and 3 one may immediately obtain the length $(r_m-l-s)-R_c$, cf. Fig. 1, of the elastic ligament between the two plastic zones as a function of the parameters involved in these figures.

5. CONCLUSIONS

The results obtained above form a sound basis for a detailed quantitative study of the cell and crack behaviour. It is a matter of an additional computational effort only to extend the figures 2-5 over sufficiently large intervals of those parameters involved in them. With such an extension and with an appropriate crack growth criterion one may almost straightforwardly derive corresponding T_m-and l-bounds for a save behaviour of such a cracked composite microcomponent. At the same time, the failure modes and the instants of their occurrence for a given cracked composite microcomponent can be predicted.

Finally, it is hoped that the very approach to the problem considered may prove itself to be a useful first step to the investigation of the interaction between longitudinal macrocracks and the near crack tip fibres in a fibrous composite.

ACKNOWLEDGEMENTS

The authors thank the Deutsche Forschungsgemeinschaft for the financial support.

REFERENCES

1. K. Herrmann, In J.W. Provan (Ed.), SM Study No. 12 "Continuum Models of Discrete Systems", University of Waterloo Press, pp. 313-338, 1978.
2. K. Herrmann, I. Mihovsky, Rozprawy Inzynierskie-Engn. Trans. 31, 2 (1983), 165-177.
3. K. Herrmann, I. Mihovsky, Mechanika Teoretyczna i Stosowana 22, 1/2 (1984), 25-39.
4. K. Herrmann, I. Mihovsky, DFG-Research Report, LTM, Paderborn University, 1987.
5. K. Herrmann, I. Mihovsky, to appear.
6. D.S. Dugdale, J. Mech. Phys. Solids 8 (1960), 100-104.
7. K. Herrmann, I. Mihovsky, ZAMM 67 (1987), T196-T197.
8. L.M. Kachanov, Foundations of the Theory of Plasticity, North-Holland, Amsterdam/London, 1971.
9. K. Herrmann, I. Mihovsky, ZAMM 68 (1988), T193-T194.
10. H.F. Bueckner, ZAMM 50 (1970), 529-546.
11. I.R. Rice, Int. J. Solids Structures 8 (1972), 751-758.
12. R.D. Gregory, Math. Proc. Camb. Phil. Soc. 81 (1977), 497-521.
13. R.D. Gregory, Math. Proc. Camb. Phil. Soc. 85 (1979), 523-538.

ON THE EFFECTIVE YOUNGS MODULUS OF ELASTICITY FOR POROUS MATERIALS

PART I: THE GENERAL MODEL EQUATION

Petrisor Mazilu
Petersenstr. 30
Institut für Umformtechnik
Technische Hochschule
D-6100 Darmstadt

Gerhard Ondracek
Mauerstr. 5
Institut für Gesteinshüttenkunde
- Glas, Keramik, Binde- und Verbundwerkstoffe -
Rheinisch-Westfälische Technische Hochschule
D-5100 Aachen

Summary

The determination of the effective moduli of elasticity represents one of the most basic tasks of materials science on composite materials nowadays. There are two concepts used to date for this determination, which are
- the *bound concept* using variational methods in which the averaged values result from the principles of the minimum of elastic potential and complementary energy.
- the *model concept* using direct methods in which the averaged stresses and distortions are calculated with the aid of Hooke's law.

Whereas the bound concept provides upper and lower bounds, between which the effective Youngs moduli of elasticity have to be expected, the model concept results in single approximate values via an effective Hooke's tensor.
Based on the model concept the present paper treats the problem of the effective elastic modulus of porous materials, considering materials with closed porosity as a limiting case of two-phase composite materials.
The derivation in this theoretical first part starts by assuming a two-phase material with matrix-type microstructure, where the two phases behave isotropically. In order to simplify the calculations it is presumed that the following proportion is valid between the elastic moduli of the inclusions (v_D, E_D) and of their matrix (v_M, E_M):

$$v_D \, E_M = v_M \, E_D$$

which, as a limiting case, exactly holds true for porous materials. - The two phase material then is subdivided into elementary cells (finite elements), where the elementary cell consists of a cube of given elastic materials in which spheroidal inclusion in any orientation are discontinuously embedded in a matrix phase. The mean stresses and strains are calculated for this elementary cell by dividing it into small, disjunct prisms. An effective modulus of elasticity is approximately calculated for each prism. The final effective modulus of elasticity is determined on the basis of a new averaging over all prisms. The resulting analytical formula for the effective Youngs modulus of elasticity depends on the elastic moduli of the phases on the phase concentration as well as on the axial ratio of the spheroid and its orientation. If in this formula the modulus for the inclusion is assumed as equal to zero, then the effective modulus of elasticity of porous materials is obtained as a function of porosity and pore structure.
The second part of the paper, published elsewhere in due course [2], is concerned with the theory of special and limiting cases of the model equation and a comparison between experimental and calculated Youngs moduli of elasticity. The comparison is made for porous glass and porous calciumtitanate ceramic and - summarizing - for experimental data of sintered metals and ceramics with spherical porosity taken from the literature.

1. General description of the model

Let a spheroid ε (rotational axis $x_3 = 2b$, secondary axis x_1, $x_2 = 2a$) be given in the space
$(x_1 = \Phi, x_2 = \Omega, x_3 = \chi)$ In this space, let a cubic elementary cell W be located (see also Fig. 1)
with the edge length l, which for its part encloses the above mentioned spheroid of revolution.
The centres of gravity of the spheroid and elementary cell coincide on the common origin of
coordinates O of the coordinate system $\Sigma = $ O, x_1, x_2, x_3 of the cube and $\Sigma' = $ O, x_1, x_2, x_3
of the spheroid (see also Fig. 2). The spheroid can take on any orientation denoted by $\underline{\alpha} = \alpha_1$,
α_2, α_3 , i. e. the rotation of the coordinate system Σ' with respect to the coordinate system Σ
about the centre gravity is determined by the angle $\underline{\alpha}$.

Let (E_D, ν_D) be the modulus of elasticity or the Poisson transversal contraction ratio within the
spheroid and (E_M, ν_M) be the corresponding values outside or in the residual volume of the
elementary cell W. In order to determine the effective modulus of elasticity in the directions
x_i (i = 1, 2, 3) of the elementary cell, the present symmetry ratios permit an exclusive
determination of the modulus of elasticity E_{eff} in the direction of the coordinate x_3.

1.2 Mathematical description of the mathematical model

Let the tensor field of the stresses $\underline{\sigma}$ be

$$\sigma = \begin{pmatrix} \sigma_{11} & \sigma_{12} & \sigma_{13} \\ \sigma_{12} & \sigma_{22} & \sigma_{23} \\ \sigma_{13} & \sigma_{23} & \sigma_{33} \end{pmatrix}$$

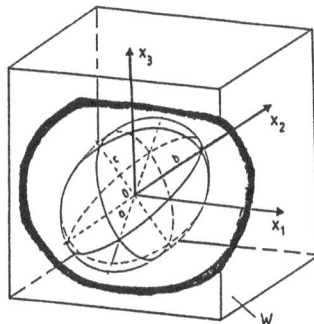

Fig. 1: Spheroid in cube

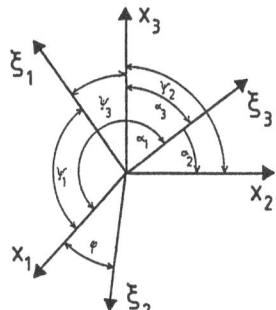

Fig. 2: Definition of the
direction angle

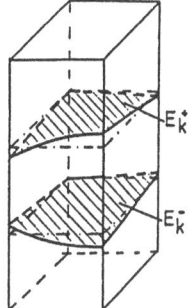

Fig. 3: Single prism

Let the vector field of the displacements $\underline{u} = (u_1, u_2, u_3)$ of the tensor field $\underline{\sigma}$ be determined in such a way that the equilibrium conditions

$$\frac{\partial\,\sigma_{ij}}{\partial\,x_j} = 0, \quad (j,i = 1, 2, 3) \tag{1}$$

and the law of change in shape for isotropic linear elastic bodies

$$\frac{\partial\,u_1}{\partial\,x_1} = \frac{1}{E_{D,A}}\left(\sigma_{11} - v_{D,A}\left(\sigma_{22} + \sigma_{33}\right)\right) \tag{2}$$

$$\text{(with permutations)}$$

and with $G = \dfrac{E}{2(1 + v)}$

$$\frac{1}{2}\left(\frac{\partial\,u_i}{\partial\,x_j} + \frac{\partial\,u_i}{\partial\,x_i}\right) = \frac{1}{2\,G_{D,M}}\,\sigma_{ij} \quad \text{for } i \neq j$$

are valid within the cell W.

The following conditions of continuity must be observed at the surface $\delta\epsilon$ of the spheroid:

$$\sigma^I_{ij}\,n_j\,|_{\partial e} = \sigma^{II}_{ij}\,n_j\,|_{\partial e}\,, \quad U^I_i\,|_{\partial c} = U^{II}_i\,|_{\partial e} \tag{3}$$

Here n_i $(i = 1, 2, 3)$ denotes the components of the normal $\underline{n}$. Furthermore, the following boundary conditions must be fulfilled on the surfaces of the cube $\partial W^{\pm}_{x_1}, \partial W^{\pm}_{x_2}, \partial W^{\pm}_{x_3}$:

$$\sigma_{11}\,|_{\partial W^{\pm}_{x_1}} = \sigma_{12}\,|_{\partial W^{\pm}_{x_2}} = \sigma_{13}\,|_{\partial W^{\pm}_{x_3}} = 0$$
$$\sigma_{12}\,|_{\partial W^{\pm}_{x_1}} = \sigma_{22}\,|_{\partial W^{\pm}_{x_2}} = \sigma_{23}\,|_{\partial W^{\pm}_{x_3}} = 0 \tag{4}$$
$$\sigma_{13}\,|_{\partial W^{\pm}_{x_1}} = \sigma_{23}\,|_{\partial W^{\pm}_{x_2}} = \sigma_{33}\,|_{\partial W^{\pm}_{x_3}} = \pm\sigma$$

Let the following be valid by definition for the effective modulus of elasticity

$$E_{eff} = \frac{1}{\ell^2}\,\sigma\!\!\iint_{\partial W_{x_3}}\frac{1}{\bar{\varepsilon}_{33}}\,dx_1\,dx_2 \quad \text{with} \quad \bar{\varepsilon}_{33} = (u_3(x_1, x_2 , \ell/2) - u_3(x_1, x_2 ,\text{-}\ell/2))/\,\ell.$$

It is not yet possible to obtain a precise solution to this problem. In order to be able to implement an approximate solution the following modified problem is substituted for the problem formulated above, with the equilibrium conditions (1) remaining unchanged: the constitutive laws (2) are replaced by the relations

$$\frac{\partial u_3}{\partial x_3} = \frac{1}{E_M}\,\sigma_{33} \quad ; \quad 0 = \sigma_{ij} \text{ for } (i, j) \neq (3,3) \tag{5}$$

outside the spheroid and

$$\frac{\partial u_1}{\partial x_1} = \frac{1}{E_D}\left(\sigma_{11} - v_D\left(\sigma_{22} + \sigma_{33}\right)\right) ; \quad \frac{1}{2}\left(\frac{\partial u_i}{\partial x_j} + \frac{\partial u_j}{\partial x_i}\right) = \frac{1}{2\,G_I}\,\sigma_{ij} \text{ for } i \neq j \tag{6}$$

(times all permutations) inside the spheroid.

The conditions of continuity (3) and the boundary conditions (4) remain unchanged.

A problem modified in this way can be described as follows from a mechanical point of view: the continuum outside the spheroid is replaced by prismatic fibres running parallel to the x_3 coordinate. However, this mechanical model also represents the borderline case of a division of the elementary cell into finite volume elements. The effective modulus of elasticity of the cell is reduced by this substitution. Let $Z_1, ..., Z_n$ be the number of disjunctive prisms filling the entire elementary cell. Those part of prisms located inside the spheroid are denoted by $Z_{K(i)}$ and those outside by $Z^{\pm}_{K(o)}$. The "+" sign denotes the upper portion of the prism and the "-" sign the lower portion.

Inside each prism the penetrating surface of the spheroid is approximated by the plane E^+_K (or E^-_K see also Fig. 3).

The original prismatic bodies $Z_{K(i)}$ and $Z^{\pm}_{K(o)}$ are thus approximated by the prismatic bodies $\overline{Z}_{K(i)}$ and $\overline{Z}^{\pm}_{K(o)}$ respectively. The position of the planes $E^{\pm}_K$ is determined in such a way that

$$\text{volume } (\overline{Z}_{K(i)}) = \text{volume } (Z_{K(i)}) \quad \text{and} \quad \text{volume } (\overline{Z}^{\pm}_{K(o)}) = \text{volume } (Z^{\pm}_{K(o)}) \qquad (7)$$

The original mathematical problem is modified as follows:
The equilibrium condition (1) remains unchanged. The constitutive laws (5) and (6) are valid in $\overline{Z}^{\pm}_{K(o)}$ with (E_N, v_M) and in $Z_{K(i)}$ with (E_D, v_D) respectively.

The conditions of continuity (3) are reduced to

$$\sigma^{(D)}_{33}|_{E_K} = \sigma^{(M)}_{33}|_{E_K} \quad \text{and} \quad u^{(D)}_3|_{E^{\pm}_K} = u^{(M)}_3|_{E^{\pm}_K} \qquad (8)$$

No continuities are assumed between the prisms $Z^{\pm}_{K(o)}$.

In order to solve the problem modified in this way, let the following homogenous state of stress be given:

$$\sigma = \begin{pmatrix} 0 & 0 & 0 \\ 0 & 0 & 0 \\ 0 & 0 & \sigma \end{pmatrix}. \qquad (9)$$

In accordance with the constitutive law (6) the distortions

$$\varepsilon_{11} = \varepsilon_{22} = -\frac{v_D}{E_D}\sigma \; ; \; \varepsilon_{33} = \frac{1}{E_D}\sigma \; ; \; \varepsilon_{12} = \varepsilon_{13} = \varepsilon_{23} = 0 \qquad (10)$$

and the displacements

$$u_1 = -\frac{v_D}{E_D}\sigma x_1 \; ; \; u_2 = -\frac{v_D}{E_D}\sigma x_2 \; ; \; u_3 = -\frac{1}{E_D}\sigma x_3 \qquad (11)$$

correspond to the stress field equation (9) within the spheroid.

Outside the spheroid, it follows from constitutive law (5) and the continuity condition (3)

$$u_3 = -\frac{1}{E_M}\sigma x_3 - \sigma\left(\frac{1}{E_M}-\frac{1}{E_D}\right)Z^\pm(x_1, x_2),$$

where $(x_1, x_2, Z^\pm(x_1, x_2))$ denotes the coordinates of the intersection of the straight lines $x_1 =$ const., $x_2 =$ const. with the upper or lower surface of the spheroid, respectively. The displacements on the cell surface $\partial W_{x_3}^\pm$ in the direction of the x_3 coordinate are then determined as

$$u_3\left(x_1, x_2, \ell/2\right) = \frac{1}{E_M}\sigma\frac{\ell}{2} - \sigma\left(\frac{1}{E_M}-\frac{1}{E_D}\right)Z^+(x_1, x_2),$$

$$u_3\left(x_1, x_2, -\ell/2\right) = -\frac{1}{E_M}\sigma\frac{\ell}{2} - \sigma\left(\frac{1}{E_M}-\frac{1}{E_D}\right)Z^-(x_1, x_2).$$

The average elongation $\qquad \overline{\varepsilon_{33}}(x_1, x_2) = \dfrac{u_3\left(x_1, x_2, \ell/2\right) - u_3\left(x_1, x_2, -\ell/2\right)}{\ell}$

is then written with $\qquad C(x_1, x_2) = \dfrac{Z^+(x_1, x_2) - Z^-(x_1, x_2)}{\ell}$

as $\qquad \overline{\varepsilon_{33}}(x_1, x_2) = \dfrac{1}{E_M}\sigma - \left(\dfrac{1}{E_M}-\dfrac{1}{E_D}\right)\sigma C(x_1, x_2)$

The effective modulus of elasticity is written with the aid of the definition

$$E_{eff} = \frac{1}{\ell^2}\sigma\iint_{\partial W_{x_3}}\frac{1}{\varepsilon}dx_1\,dx_2 \quad \text{as} \quad E_{eff} = \frac{1}{\ell^2}\int_{-\ell/2}^{\ell/2}dx_1\int_{-\ell/2}^{\ell/2}\frac{dx_2}{\frac{1}{E_M}-\left(\frac{1}{E_M}-\frac{1}{E_D}\right)C(x_1, x_2)} \quad (12)$$

Let the homogenous state of stress (9) be given once again to determine the coefficient $C(x_1,x_2)$.

Within the region set of all prisms $Z_{K(i)}$, the formulae (10) and (11) are valid for the strains and displacements. Outside this region set, the strains are determined by

$$\varepsilon_{11}=\varepsilon_{22}=-\frac{v_M}{E_M}\sigma \quad ; \quad \varepsilon_{33}=\frac{1}{E_M}\sigma \quad ; \quad \varepsilon_{12}=\varepsilon_{23}=\varepsilon_{13}=0. \qquad (13)$$

The corresponding displacements thus follow as

$$u_1 = -\frac{v_M}{E_M}\sigma x_1 \quad ; \quad u_2 = -\frac{v_M}{E_M}\sigma x_2 \quad ; \quad u_3 = \frac{1}{E_M}\sigma x_3 + \sigma\left(\frac{1}{E_D}-\frac{1}{E_M}\right)\overline{Z}_{K(o)}^\pm$$

where $\overrightarrow{Z}_{K(o)}^\pm$ denotes the coordinate of the upper or lower plane $E_K^\pm$. The mean elongation of a cylinder is formulated with

$$\overline{\varepsilon_{33}} = \frac{u_3\,(\ell/2) - u_3\,(-\ell/2)}{\ell} \qquad \text{as} \qquad \overline{\varepsilon}_{33} = \frac{1}{E_M}\,\sigma + \left(\frac{1}{E_D} - \frac{1}{E_M}\right)\sigma C_K$$

Furthermore, with $\qquad C_K = \dfrac{Z_K^+ - Z_K^-}{\ell}$

the effective modulus of elasticity follows as

$$\left(\frac{1}{E}\right)_{\text{eff}} = \frac{1}{\ell^2}\sum_{K=1}^{n}\left[\frac{1}{E_M} + \left(\frac{1}{E_D} - \frac{1}{E_M}\right)C_K\right]S(E_K)\,; \qquad E_{\text{eff}} = \frac{1}{\ell^2}\sum_{K=1}^{n}\frac{S(E_K)}{\frac{1}{E_M} + \left(\frac{1}{E_D} - \frac{1}{E_M}\right)C_K}\quad(14)$$

where $S(E_K)$ denotes the surface of the plane E_K. The coefficients C_K and the areas $S(E_K)$ must therefore first be calculated in order to determine the effective modulus of elasticity. The relation

$$C_K = \frac{Z_K^+ - Z_K^-}{\ell} = \frac{\text{vol}\,(Z_{K(i)})}{\ell S(E_K)} \qquad\qquad (15)$$

follows from equation (7).

The volumes vol $(Z_{K(i)})$, the surfaces $S_{(E_K)}$ and from them the coefficients C_K can be determined as follows for a special type of cell division. A special system of elliptical coordinates is defined in order to be able to describe this particular, discrete division of the elementary cell.

As already mentioned and as can also be seen from Fig. 2, the coordinate system $\Sigma' = \{0, \xi_1 \; \xi_2, \xi_3\}$ with respect to the coordinate system $\Sigma = \{0, x_1, x_2, x_3\}$ is described as follows with the aid of the direction angle.

$$\overrightarrow{\xi_1}(\cos\psi_1, \cos\psi_2, \cos\psi_3)\;;\; \overrightarrow{\xi_2}(\cos\phi, \sin\phi, 0)\;;\; \overrightarrow{\xi_3}(\cos\alpha_1, \cos\alpha_2, \cos\alpha_3)\quad(16)$$

The restrictions $\qquad -\dfrac{\pi}{2} \le \psi \le \dfrac{\pi}{2}\;;\qquad \cos^2\alpha_1 + \cos^2\alpha_2 \neq 0 \qquad\qquad (17)$

are introduced for the further formulation without any limitation of the general orientation of the spheroid.

In the case $\alpha_1 = \alpha_2 = \dfrac{\pi}{2}$, the trivial case $\Sigma' \equiv \Sigma$ is valid. The angles ψ, ψ_1, ψ_2 and ψ_3 are now determined. If one takes into consideration the fact that $\overrightarrow{\xi_1} = \overrightarrow{\xi_2} \times \overrightarrow{\xi_3}$ then the following is obtained

$$\cos\psi_1 = \sin\Phi\,\cos\alpha_3;\; \cos\psi_2 = -\cos\Phi\,\cos\alpha_3$$

and

$$\cos\psi_3 = -\sin\Phi\,\cos\alpha_1 + \cos\Phi\,\cos\alpha_2\,.$$

In addition it follows with $\overrightarrow{\xi_3} \perp \overrightarrow{\xi_2}$

$$\cos\Phi\,\cos\alpha_1 + \sin\Phi\,\cos\alpha_2 = 0 \qquad\qquad (18)$$

from which the following is obtained

$$\sin\phi = -\frac{\cos\alpha_1}{\sqrt{\cos^2\alpha_1 + \cos^2\alpha_2}} \; ; \cos\phi = -\frac{\cos\alpha_2}{\sqrt{\cos^2\alpha_1 + \cos^2\alpha_2}} \qquad (19)$$

If a general formulation is applied

$$x_i = \bar{n}_{iK}\xi_K; \xi_i = \bar{n}_{iK}x_K \; ; \; x = \bar{N}\xi \; ; \; \xi = \bar{N}x \qquad (20)$$

then the transformation matrix N follows with equation (16) and equation (18) as

$$N = \begin{pmatrix} \sin\Phi\cos\alpha_3 & -\cos\Phi\cos\alpha_3 & -\sin\Phi\cos\alpha_1 + \cos\Phi\cos\alpha_2 \\ \cos\Phi & \sin\Phi & 0 \\ \cos\alpha_1 & \cos\alpha_2 & \cos\alpha_3 \end{pmatrix}$$

or with $N^{-1} = N^T = \bar{N}$ $\qquad (21)$

$$\bar{N} = \begin{pmatrix} \sin\Phi\cos\alpha_3 & \cos\Phi & \cos\alpha_1 \\ -\cos\Phi\cos\alpha_3 & \sin\Phi & \cos\alpha_2 \\ -\sin\Phi\cos\alpha_1 + \cos\Phi\cos\alpha_2 & 0 & \cos\alpha_3 \end{pmatrix}$$

The equation for the spheroid of revolution is in ξ_1, ξ_2, ξ_3 with the principal axes a and b (see also Fig. 2.)

$$\frac{1}{a^2}(\xi_1^2 + \xi_2^2) + \frac{1}{b^2}\xi_3^2 = 1. \qquad (22)$$

After several computing operations the desired formulation of the spheroid equation (22) is obtained with the aid of relations (19), (20), (21)

$$\left[\frac{1}{a^2} + \left(\frac{1}{b^2} - \frac{1}{a^2}\right)\cos^2\alpha_1\right]x_1^2 + \left[\frac{1}{a^2} + \left(\frac{1}{b^2} - \frac{1}{a^2}\right)\cos^2\alpha_2\right]x_2^2 +$$

$$+ \left[\frac{1}{a^2} + \left(\frac{1}{b^2} - \frac{1}{a^2}\right)\cos^2\alpha_3\right]x_3^2 + 2\left(\frac{1}{b^2} - \frac{1}{a^2}\right)\cos\alpha_1\cos\alpha_2\, x_1 x_2 + \qquad (23)$$

$$+ 2\left(\frac{1}{b^2} - \frac{1}{a^2}\right)\cos\alpha_1\cos\alpha_3\, x_1 x_3 + 2\left(\frac{1}{b^2} - \frac{1}{a^2}\right)\cos\alpha_2\cos\alpha_3\, x_2 x_3 = 1$$

With $x_1 = x, x_2 = y, x_3 = z$ and the expressions

$$A = \frac{1}{a^2} + \left(\frac{1}{b^2} - \frac{1}{a^2}\right)\cos^2\alpha_3 = \alpha$$

$$B(x,y) = \left(\frac{1}{b^2} - \frac{1}{a^2}\right)(\cos\alpha_1\cos\alpha_3\, x + \cos\alpha_2\cos\alpha_3\, y) = \beta_1 x + \beta_2 y \qquad (24)$$

$$C(x,y) = \left[\frac{1}{a^2} + \left(\frac{1}{b^2} - \frac{1}{a^2}\right)\cos^2\alpha_1\right]x^2 + \left[\frac{1}{a^2} + \left(\frac{1}{b^2} - \frac{1}{a^2}\right)\cos^2\alpha_2\right]y^2$$

$$+ 2\left(\frac{1}{b^2} - \frac{1}{a^2}\right)\cos\alpha_1\cos\alpha_2\, x\, y - 1$$

$$= \gamma_1 x^2 + \gamma_2 y^2 + 2\,\gamma_3 xy - 1$$

the spheroid equation becomes

$$A z^2 + 2 B(x, y) z + C(x, y) = 0 \qquad (25)$$

The solution of the quadratic equation for z provides

$$z_{1,2} = \frac{-B(x,y) \pm \sqrt{B^2(x,y) - A\,C(x,y)}}{A} \tag{26}$$

The z coordinates of the intersection between a normal to the x-y plane and the spheroid can thus be directly determined for all possible points (x, y). However, the description of the projection ellipse in the principal axis system $(\bar{x}, \bar{y})$ is of advantage. This coordinate system $(\bar{x}, \bar{y})$ situated in the x-y plane is rotated by the angle ψ' with respect to the x-y system. ψ' is determined by the eigenvalues of the corresponding quadratic form of the projected spheroid. To this end, the discriminent in equation (26) is put equal to zero, thus

$$B^2 - AC = 0 \tag{27}$$

With equation (24) and

$$p = -\frac{\left(\beta_1^2 - \alpha\,\gamma_1\right)}{\alpha} \quad ; \quad g = -\frac{\left(\beta_1\,\beta_2 - \alpha\,\gamma_3\right)}{\alpha} \quad ; \quad r = -\frac{\left(\beta_2^2 - \alpha\,\gamma_2\right)}{\alpha} \tag{28}$$

it follows from equation (27) $\qquad\qquad px^2 + 2\,gxy + ry^2 = 1 \tag{29}$

The eigenvalues for the quadratic form equation (29) are determined as

$$\lambda_{1,2} = \frac{p + r \pm \sqrt{(p-r)^2 + 4\,g^2}}{2} \tag{30}$$

and the resulting principal axes in the $(\bar{x}, \bar{y})$ system as $\left(\sqrt{\frac{1}{\lambda_1}}, \sqrt{\frac{1}{\lambda_2}}\right)$.

The associated angle follows from $\quad \tan 2\,\phi' = \dfrac{2\,g}{p - r} \tag{31}$

The principal axes of the ellipse in Fig. 4 follow from

$$\lambda_1 \bar{x}^2 + \lambda_2 \bar{y}^2 = 1 \qquad\qquad \text{as} \qquad\qquad \sqrt{\frac{1}{\lambda_1}} \text{ and } \sqrt{\frac{1}{\lambda_2}}. \tag{32}$$

If the coordinate system shown in Fig. 5 is now introduced then - starting from the discretization of the projection ellipse - the discrete volumes of the spheroid can be derived:

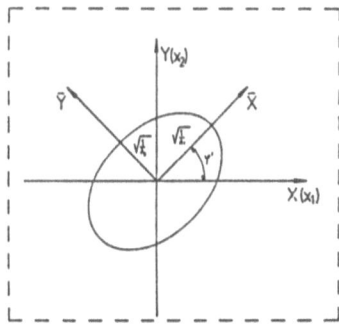

Fig. 4: Projection of the spheroid
on the x-y-plane

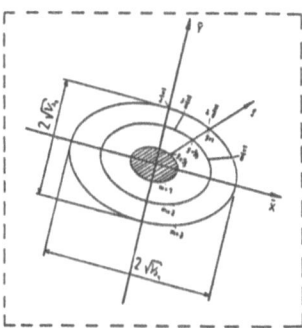

Fig. 5: Example of discretization

Let the following be valid with the parameter Θ

$$\bar{x} = \sqrt{\frac{1}{\lambda_1}}\,\rho\cos\Theta \quad ; \quad \bar{y} = \sqrt{\frac{1}{\lambda_2}}\,\rho\sin\Theta \tag{33}$$

Furthermore the following is valid

$$0 \le \rho \le 1 \; ; \; \rho_m = \frac{m}{M}\,(m = 1, ..., M)\,,$$
$$0 \le \Theta \le 2\pi \; ; \; \Theta_n^m = \frac{2\pi n}{N_m}\,(n = 1, ..., N_m)\,, \tag{34}$$
$$\rho_0 = 0 \; ; \; \Theta_0^m = 0 \; ; \; V_n^m = V(\Delta_{mn})\,.$$

With equation (26) the volume of a cylinder with the base Δ_{mn} within the spheroid becomes

$$V_n^m = \frac{2}{A}\iint_{\Delta_{mn}}\sqrt{B^2(x,y) - AC(x,y)}\;dx\,dy\,. \tag{35}$$

With equations (24), (28) and (32) it follows from equation (35)

$$V_n^m = \frac{2}{\sqrt{A}}\iint_{\Delta_{mn}}\sqrt{1 - (\lambda_1\bar{x}^2 + \lambda_2\bar{y}^2)}\;d\bar{x}\,d\bar{y}\,. \tag{36}$$

with equation (33) it follows

$$d\bar{x}\,d\bar{y} = \frac{1}{\sqrt{\lambda_1}\,\sqrt{\lambda_2}}\,\rho\,d\rho\,d\psi \tag{37}$$

and thus from equation (36) one finds

$$V_n^m = \frac{2}{\sqrt{A}\,\sqrt{\lambda_1\lambda_2}}\iint_{\Delta_{mn}}\rho\sqrt{1 - \rho^2}\;d\rho\,d\Theta = \frac{2\left(\frac{m}{n} - \frac{m}{n-1}\right)}{\sqrt{A}\,\sqrt{\lambda_1\lambda_2}}\int_{\rho_{m-1}}^{\rho_m}\rho\sqrt{1 - \rho^2}\;d\rho\,. \tag{38}$$

After a further integration, the partial volume follows with equation (34)

$$V_n^m = \frac{2}{3}\frac{2}{N_m}\frac{\sqrt{\left(1 - \left(\frac{m-1}{M}\right)^3\right)} - \sqrt{\left(1 - \left(\frac{m}{M}\right)^2\right)^3}}{\sqrt{A}\,\sqrt{\lambda_1\lambda_2}}\,. \tag{39}$$

If one now considers a special example with $N_1 = 1$, $N_2 = 8$, $N_3 = 16$ and $M = 3$, then the partial volumes are calculated as

$$V_1^1 = \frac{4}{3}\,\pi\,\frac{0.161}{\sqrt{A}\,\sqrt{\lambda_1\lambda_2}} \; ; \; V_{1\text{-}8}^2 = \frac{\pi}{6}\,\frac{0.42}{\sqrt{A}\,\sqrt{\lambda_1\lambda_2}} \; ; \; V_{1\text{-}16}^3 = \frac{\pi}{12}\,\frac{0.437}{\sqrt{A}\,\sqrt{\lambda_1\lambda_2}} \tag{40}$$

In order to determine the coefficient C_K equation (15), it only remains to determine the surface $S(\Delta_{mn}) = S_n^m$ of the partial cylinder in the x-y plane (see also Fig. 5). The following is valid with the functional determinant in equation (37)

$$S_n^m = \iint_{\Delta_{mn}}dx\,dy = \frac{1}{\sqrt{\lambda_1\lambda_2}}\int_{\rho_{m-1}}^{\rho_m}\rho\,d\rho\int_{\Theta_{n-1}^m}^{\Theta_n^m}d\theta = \frac{1}{\sqrt{\lambda_1}\,\sqrt{\lambda_2}}\left(\Theta_n^m - \Theta_{n-1}^m\right)\left(\rho_m^2 - \rho_{m-1}^2\right), \tag{41}$$

or with equation (34)

$$S_n^m = \frac{1}{n}\frac{1}{\sqrt{\lambda_1\lambda_2}}\frac{2\pi}{N\,m}\left[\left(\frac{m}{M}\right)^2 - \left(\frac{m-1}{M}\right)^2\right]\,. \tag{42}$$

It therefore follows in accordance with relation (15) to the volume constancy with equation (39) and equation (42)

$$c_{\Pi}^m = \frac{4}{3} \frac{\sqrt{\left(1 - \left(\frac{m-1}{M}\right)^2\right)^3} - \sqrt{\left(1 - \left(\frac{m}{M}\right)^2\right)^3}}{\sqrt{A}\left[\left(\frac{m}{M}\right)^2 - \left(\frac{m-1}{M}\right)^2\right] \ell}$$ (43)

If one then returns to the example given above then the following values can be calculated from equation (43)

$$c_1^1 \approx 1.992 \frac{1}{\sqrt{A} \ \ell} \quad ; \quad c_{1-8}^2 = 1.68 \frac{1}{\sqrt{A} \ \ell} \quad ; \quad c_{1-16}^3 = 1.04 \frac{1}{\sqrt{A} \ \ell}$$ (44)

According to relation (14) it thus follows for the effective modulus of elasticity

$$E_{eff} = \frac{E_D E_M}{\ell^2}\left[\frac{\ell^2 - \frac{\pi}{\sqrt{\lambda_1 \lambda_2}}}{E_D} + \sum_{m=1}^{M}\sum_{n=1}^{N_m}\frac{S_n^m}{E_D + c_{\Pi}^m \ (E_M - E_D)}\right].$$ (45)

3.   General model equation for porous materials

With the relations (40) and (44) contained in the example, the following is obtained from relation (45) extended by "ℓ" (discrete distribution M = 3, $N_1 = 1$, $N_2 = 8$, $N_3 = 16$)

$$\frac{E_{eff}}{E_M} = 1 - \frac{\pi}{\ell^2 \sqrt{\lambda_1 \lambda_2}}\left[1 - \frac{1}{9\left(1 + \frac{1.99}{\ell\sqrt{A}}\left(\frac{E_M}{E_D} - 1\right)\right)} - \frac{1}{3\left(1 + \frac{1.68}{\ell\sqrt{A}}\left(\frac{E_M}{E_D} - 1\right)\right)} - \frac{1}{\frac{9}{5}\left(1 + \frac{1.04}{\ell\sqrt{A}}\left(\frac{E_M}{E_D} - 1\right)\right)}\right]$$ (46)

The equation shows that the normalized effective elasticity modulus E_{eff}/E_M is a function of E_M/E_D but not of an absolute value (E_M or E_D).

Two expressions contain the microstructure data: $\ell\sqrt{A}$ and $\ell^2\sqrt{\lambda_1 \lambda_2}$:

$$\ell\sqrt{A} = \frac{\ell}{a}\sqrt{1 + \left(\frac{a^2}{b^2} - 1\right)\cos^2\alpha_3}$$ (47a)

from which with $\frac{a}{b} = \frac{x}{z} = k$ and $\frac{4\pi a^2 b}{3\ell^3}$ it follows

$$\ell\sqrt{A} = \frac{\sqrt[3]{\frac{4\pi}{3 \ c_D}}\sqrt{1 + \left[\left(\frac{x}{z}\right)^2 - 1\right]\cos^2\alpha_3}}{\sqrt{\frac{x}{z}}},$$ (47b)

$$\ell^2\sqrt{\lambda_1 \lambda_2} = \sqrt{\left(\frac{\ell^2\beta_1^2}{\alpha} - \ell^2\gamma_1\right)\left(\frac{\ell^2\beta_2^2}{\alpha} - \ell^2\gamma_2\right) - \left(\frac{\ell^2\beta_1\beta_2}{\alpha} - \ell^2\gamma_3\right)^2}$$ (48a)

with

$$\frac{\ell^2\beta_1^2}{\alpha} = \left(\frac{4\pi}{3c_D}\right)^{2/3} k^{-2/3}\frac{(k^2 - 1)\cos^2\alpha_1 \cos^2\alpha_3}{1 + (k^2 - 1)\cos^2\alpha_3},$$ (49a)

$$\ell^2\gamma_1 = \left(\frac{4\pi}{3c_D}\right)^{2/3} k^{-2/3}\left[1 + (k^2 - 1)\cos^2\alpha_1\right],\tag{49b}$$

$$\frac{\ell^2\beta_2^2}{\alpha} = \left(\frac{4\pi}{3c_D}\right)^{2/3} k^{-2/3}\frac{(k^2 - 1)^2 \cos^2\alpha_2 \cos^2\alpha_3}{1 + (k^2 - 1)\cos^2\alpha_3},\tag{49c}$$

$$\ell^2\gamma_2 = \left(\frac{4\pi}{3c_D}\right)^{2/3} k^{-2/3}\left[1 + (k^2 - 1)\cos^2\alpha_2\right],\tag{49d}$$

$$\frac{\ell^2\beta_1\beta_2}{\alpha} = \left(\frac{4\pi}{3c_D}\right)^{2/3} k^{-2/3}\frac{(k^2 - 1)^2 \cos\alpha_1 \cos\alpha_2 \cos^2\alpha_3}{1 + (k^2 - 1)\cos^2\alpha_3},\tag{49e}$$

$$\ell^2\gamma_3 = \left(\frac{4\pi}{3c_D}\right)^{2/3} k^{-2/3} (k^2 - 1)\cos\alpha_1 \cos\alpha_2,\tag{49f}$$

where $k = \dfrac{a}{b}$ is valid and the concentration is denoted by $c_D = \dfrac{4\,a^2 b}{3\,\ell^3}$

Substituting equation (49) in equation (48a) results in

$$\ell^2\sqrt{\lambda_1\lambda_2} = \left(\frac{4\pi}{3c_D}\right)^{2/3} k^{-2/3} \left[\left[\frac{(k^2 - 1)^2 \cos^2\alpha_1 \cos^2\alpha_3}{1 + (k^2 - 1)\cos^2\alpha_3} - 1 - (k^2 - 1)\cos^2\alpha_1\right] \left[\frac{(k^2 - 1)^2 \cos^2\alpha_2 \cos^2\alpha_3}{1 + (k^2 - 1)\cos^2\alpha_3} - 1 - (k^2 - 1)\cos^2\alpha_2\right] - \left[\frac{(k^2 - 1)^2 \cos\alpha_1 \cos\alpha_2 \cos^2\alpha_3}{1 + (k^2 - 1)\cos^2\alpha_3} - (k^2 - 1)\cos\alpha_1 \cos\alpha_2\right]^2\right]^{-\frac{1}{2}}\tag{48b}$$

If one designates

$$\frac{1}{k^2 - 1} = \chi\tag{50}$$

and

$$H = (k^2 - 1)^2\left[\frac{\cos^2\alpha_1\cos^2\alpha_3}{\chi + \cos^2\alpha_3} - (\chi + \cos^2\alpha_1)\right]\left(\frac{\cos^2\alpha_2\cos^2\alpha_3}{\chi + \cos^2\alpha_3} - (\chi + \cos^2\alpha_2)\right) - \left(\frac{\cos^2\alpha_1\cos^2\alpha_2\cos^2\alpha_3}{\chi + \cos^2\alpha_3} - \cos\alpha_1\,\cos\alpha_2\right)^2\right]\tag{51a}$$

then the following is valid

$$\ell^2\sqrt{\lambda_1\lambda_2} = \left(\frac{4\pi}{3\,c_D\,k}\right)^2 \sqrt{H}.\tag{48c}$$

Equation (51a) can be transformed further:

$$H = (k^2 - 1)^2\left\{(\chi + \cos^2\alpha_1)(\chi + \cos^2\alpha_2) - \cos^2\alpha_1\cos^2\alpha_2 - \frac{R}{(\chi + \cos^2\alpha_3)^2}\right\},$$

$$R = \cos^2\alpha_1\cos^2\alpha_3\,(\chi + \cos^2\alpha_2) + \cos^2\alpha_2\cos^2\alpha_3\,(\chi + \cos^2\alpha_1) - 2\cos^2\alpha_1\cos^2\alpha_2\cos^2\alpha_3,\tag{51b}$$

from which it follows

$$H = -\frac{(k^2 - 1)^2}{\chi + \cos^2\alpha_3}\left\{(1 - \cos^2\alpha_3)\cos^2\alpha_3 + (k^2 - 1)^2\left[\chi^2 + \chi(1 - \cos^2\alpha_3)\right]\right\} =$$

$$= - \frac{(k^2 - 1)^2}{\chi + \cos^2\alpha_3} \left\{ (1 - \cos^2\alpha_3) \cos^2\alpha_3 + (k^2 - 1)[\chi + 1 - \cos^2\alpha_3] \right\} =$$

$$= (k^2 - 1) + (1 - \cos^2\alpha_3) - \frac{(1 - \cos^2\alpha_3) \cos^2\alpha_3}{\chi + \cos^2\alpha_3} =$$

$$= (k^2 - 1) + (1 - \cos^2\alpha_3)\left(1 - \frac{\cos^2\alpha_3}{\chi + \cos^2\alpha_3}\right) =$$

$$= (k^2 - 1) + \frac{(1 - \cos^2\alpha_3)}{\chi + \cos^2\alpha_3} = 1 + \frac{(1 - \cos^2\alpha_3)}{\chi + \cos^2\alpha_3} = \frac{\chi + 1}{\chi + \cos^2\alpha_3} \qquad (51c)$$

or

$$H = \frac{\dfrac{1}{k^2 - 1} + 1}{\dfrac{1}{k^2 - 1} - \cos^2\alpha_3} = \frac{k^2}{1 + (k^2 - 1)\cos^2\alpha_3} \qquad (51d)$$

and thus

$$\ell^2\sqrt{\lambda_1\lambda_2} = \sqrt[3]{\left(\frac{4\pi}{3\,c_D\,k}\right)^2} \frac{k}{\sqrt{1 + (k^2 - 1)\cos^2\alpha_3}} =$$

$$= \frac{k\sqrt[3]{\left(\dfrac{4\pi}{3\,c_D\,k}\right)^2}}{\sqrt{1 + (k^2 - 1)\cos^2\alpha_3}} = \frac{\sqrt[3]{k\left(\dfrac{4\pi}{3\,c_D}\right)^2}}{\sqrt{1 + (k^2 - 1)\cos^2\alpha_3}}, \qquad (48d)$$

or

$$\ell^2\sqrt{\lambda_1\lambda_2} = \frac{\dfrac{x}{z}\sqrt[3]{\left(\dfrac{4\pi}{3\,c_D}\dfrac{z}{x}\right)^2}}{\sqrt{1 + \left(\left[\dfrac{x}{z}\right]^2 - 1\right)\cos^2\alpha_3}} = \frac{\sqrt[3]{\dfrac{x}{z}\left(\dfrac{4\pi}{3\,c_D}\right)^2}}{\sqrt{1 + \left(\left[\dfrac{x}{z}\right]^2 - 1\right)\cos^2\alpha_3}} \qquad (48e)$$

and

$$\ell\sqrt{A} = \frac{\sqrt[3]{\left(\dfrac{4\pi}{3\,c_D}\right)}\sqrt{1 + \left(\left[\dfrac{x}{z}\right]^2 - 1\right)\cos^2\alpha_3}}{\sqrt[3]{\dfrac{x}{z}}}. \qquad (47b)$$

The pore case (spheroidal pores) corresponds to the borderline case $(E_M/E_D) \to \infty$. It then follows from equations (46b), (47b) and (48e)

$$E_{eff} = E_M\left(1 - \frac{\pi}{\ell^2\sqrt{\lambda_1\lambda_2}}\right) = E_M\left(1 - \frac{\pi\left|\sqrt{1 + \left(\left[\dfrac{x}{z}\right]^2 - 1\right)\cos^2\alpha_3}\right|}{\sqrt[3]{\dfrac{x}{z}\left(\dfrac{4\pi}{3\,c_D}\right)^2}}\right). \qquad (52)$$

Since the spheroid of revolution cannot completely fill up the space, equation (46b) is not valid throughout the entire concentration range. The maximum concentration depends on the form and orientation of the rearranged phases. The following is valid, for example,

- for spheres $c_D = 0.52$
- for directed cylinders (cylinder axis parallel to the cube edge of the finite element) $c_D = 0.78$

In treating the microstructure-field property relationship, the validity of the model equation was achieved by the principle of schematic differentiation [1], which however in the present case fails due to mathematical difficulties (c_D with exponent cannot be treated as a differential). This is, why the problem has to be solved approximately by a first, engineering approach: for $c_D = 1$ the normalized effective Youngs modulus of elasticity for porous materials should be zero ($\frac{E_{eff}}{E_M} = E_p = 0$), but equation 52 provides

$$\frac{E_{eff}}{E_M} = 1 - 1.21 \frac{\sqrt{1 + \left(\left[\frac{x}{z}\right]^2 - 1\right) \cos^2 \alpha_3}}{\sqrt[3]{\frac{x}{z}}} \cdot c_D{}^{3/2} \tag{52a}$$

Substracting this term from equ. 52 and observing,

- that this "compensation" should reduce strongly according to experimental results with decreasing porosity

- that the exponent $\frac{3}{2}$ is offered for the porosity by the derivation of equ. (52).

the model equation for porous materials as a I. engineering approach follows as

$$\frac{E_{eff}}{E_M} = 1 - 1.21 \frac{\sqrt{1 + \left(\left[\frac{x}{z}\right]^2 - 1\right) \cos^2 \alpha_3}}{\sqrt[3]{\frac{x}{z}\left[\frac{1}{c_D}\right]^2}} - c_D{}^{3/2} \left[1 - 1.21 \frac{\sqrt{1 + \left(\left[\frac{x}{z}\right]^2 - 1\right) \cos^2 \alpha_3}}{\sqrt[3]{\frac{x}{z}}} \right] \tag{52b}$$

Limiting cases for this equation as spherical, needle and disc shaped pores, its convergency with the bound equations and the comparison of calculated and measured Youngs moduli for porous materials are treated in part II of the paper [2].

References
[1] Ondracek G., Reviews on Powder Metallurgy and Physical Ceramics 3-3/4 (1987) 205
[2] Mazilu P., Ondracek G., Int. Conf. on the Mechanics, Physics and Structure of Materials, Thessaloniki August (1990) in print

Acknowledgement
The worked reported in the present contribution was sponsored by the Deutsche Forschungsgemeinschaft (DFG) and the Internationales Büro des Bundesministeriums für Forschung und Technologie im Forschungszentrum Jülich (KFA-INT). Typing and drafting was done by Mrs. Monika Schmitz, Mrs. Monika Schüller, Mr. Aldo Boccaccini and Mr. Joannis Trechas. The authors gratefully appreciate this assistance.

THE CRACK DRIVING FORCE IN DAMAGED PARTICLE STRENGTHENED CERAMICS

F.E. Buresch

Institut für Computeranwendungen, Universität Stuttgart
Pfaffenwaldring 27, D-7000 Stuttgart 80

1 Scope of the presentation

Due to the anisotropic and inhomogeneous microstructure of brittle materials such as ceramics, failure phenomena in this class of materials are attributed to elastic damage caused by stress induced microcracking. Generally stress induced microcracks develop from microcrack nuclei, specifically in the regions of high macroscopic stress concentration of the elastic stress field ahead of a critical crack or notch. Microcrack nuclei are characterized by specific microscopic stress intensity factors (MSIF), which are induced by residual stresses of the second kind of microstructural features, such as pores or triple points of grain boundary junctions. These residual stresses are mainly due to thermal expansion, mismatch of grains, phases in facets, or to phase transformations. The magnitude of the MSIF depends on the square root of grain boundary facet length, which are statistically distributed in the volume of the microstructure, and can be experimentally quantified with a quantitative microstructural analysis.

Corresponding to the nature of the MISF's of the nuclei, their statistical distribution, the volume, the loading, and the environmental conditions, a more or less anisotropic microcrack field develops in a process zone ahead of a macroscopic stress concentration during monotonic loading above the elastic limit. Due to the microcrack porosity, which is characteristic for elastic damage, the dilated process zone is equivalent to a material with different mechanical properties, with respect to the safe undamaged surrounding material.

In previous works it was shown [1,2] that at a critical microcrack configuration some favorable oriented microcracks join together to or with a macrocrack, which induces local unloading of the process zone. This instability criterion is equivalent to a critical strain energy density of the damage zone. Thus, monotonic loading of a component causes subsequent unloading, reloading processes of the crack region, and a quantized macrocrack growth in discrete steps. This is experimentally observed by different methods [3,4].

In this work we will give additional theoretical and experimental contributions to previous results. This specifically concerns the influence of residual stresses of the first kind of the dilated process zone, which are a consequence of the quantized crack growth. It will be quantitatively shown that the release rate of the residual elastic strain energy of the process zone, governs the crack resistance of this class of materials during crack growth.

2   Elastic damage in a nonlinear deformed beam.

2.1   Influence of elastic damage on the elastic and inelastic strain energy during bending.

From numerous experimental observations it is obvious [3-6] that a beam of a ceramic, typically for alumina loaded by an end couple, shows a nonlinear nonelastic response. Thus, a bend beam of a DCB specimen goes partially inelastic, and the applied moment along the beam, uniformly produce an elastic core of height ζh on either side of the neutral axis, and an inelastic outer layer $(1-\zeta)$h (fig. 1). The free length of the beam is equivalent to a crack of length a. In partially elastic damage materials, the elastic parameters in the elastically deformed core are different than that of the inelastically deformed outer layers.

As mentioned, the inelastic deformation is due to stress induced microcracking, which arises locally, if the superposition of the microscopic stress intensity with the macroscopic one reach critical levels. This will first happen during loading in front of a critical microcrack nuclei with the highest MISF's, however, the largest grain facets break first. The damage process continues with increasing load, and more microcrack nuclei with decreasing MSIF are involved in the micro-fracture process. This process is finished at a critical load, if the microcrack configuration reaches a critical situation, where some favorable oriented microcracks join together, and the energy density in the damaged region is extreme. At this mechanical state the macrocrack grows. From experimental and theoretical observations it can be concluded, that the critical volume density of microcracks is somewhere in the range of 20% [1,7]. This is nearly equivalent to a density of a row of parallel oriented cracks of 50%. It is obvious that elastic damage changes the elastic properties in the same magnitude, and is assumed to be homogeneous in the damage zone. Thus, an elastic mismatch arises at the boundary of the damage zone.

The mechanical problem was evaluated for a glued, plastically damaged cantilever beam by A.G. Atkins and J.W. Mai [8]. It will be extended in this work with some constitutive modifications for an elastic microcracking material.

The stiffness of the elastic damaged beam is not the same as the fully elastic beam, due to the fact that damage in the outer layer change the volume. However, despite the fact that Hook's law did not hold, it is assumed that plane sections remain plane, and the stress strain diagram of the damaged layers of the beam for tension and compression are the same, that is, the neutral axis of the beam crosses the center of gravity. The specific microstructural processes characterize the elastic limit of the material.

Due to the fact that the Young's modulus of the outer microcracked layer changes from E to E_m, the rotation of the elastic and inelastic deformed sections of the beam are not the same, and unloading/loading lines are not parallel to the initial stiffness line OY. The work area OYGF represents irreversible inelastic work U_{diss}, required to build up the microcracked zones in the beam of constant length a, to the level given by the appropriate P or M, and in addition it will be shown that area OYGF contains residual elastic strain energy as well (fig. 2).

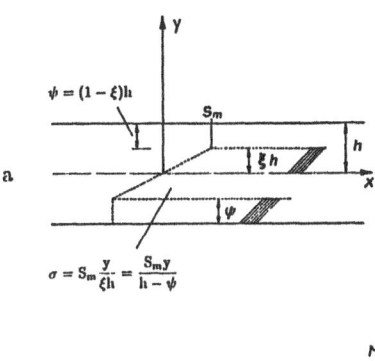

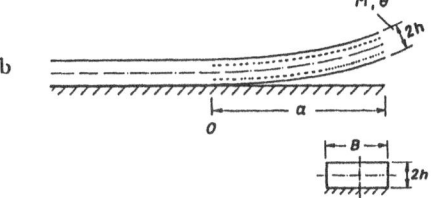

Figure 1: Deformation of an elasto-plastic beam
a   Stress distribution across the beam depth with the regions of elastic and plastic bending
b   Rotation of an elasto-plastic beam

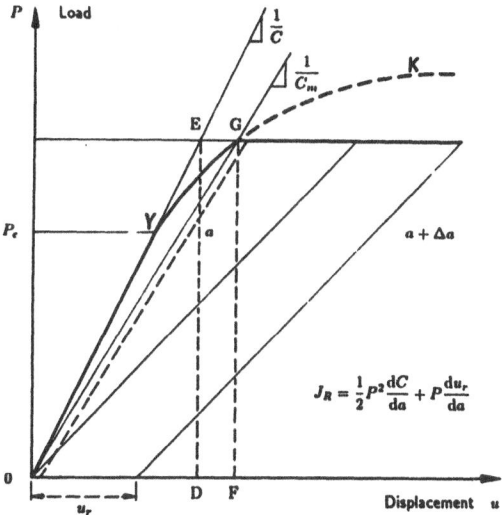

Figure 2: Load-displacement curve with incremental crack growth Δa at constant load

To get a material specific formula for failure of the beam, we will look for the strain energy densities, which, during inelastic deformation of the beam, contribute to the energy release rate during fracture. This is different for the propagation of a crack without elastic damage. We will identify the damage model with a brittle material such as a coarse grained ceramic, which shows stress induced microcracking during monotonic loading, beginning at and beyond the elastic limit.

In the loaded beam, a inhomogeneous elastic strain energy density develops in the pure elastically deformed core, and with the notation of fig. 1, is given as

$$\frac{S_m^2}{2E} \frac{y^2}{\zeta^2 h^2} \tag{1}$$

whereas, in the nonlinear deformed microcracked outer layer, the elastic strain energy is constant at

$$\frac{S_m^2}{2E_m} \tag{2}$$

With these formulations it is assumed, that constant homogeneous cohesion strength S_m is valid for the boundaries between the elastically and inelastically deformed zones of a beam with Young's moduli E and $E_m \leq E$, respectively. The reduced Young's modulus is assumed to be homogeneous in the inelastically deformed outer layers of thickness $\psi = (1-\zeta)h$, where 2h is the height of the beam. Then, with a constant cohesion strength of S_m of the material, the moment of the beam is given [8] by

$$M_y = \frac{Bh^2 S_m}{3}(3 - \xi^2) \tag{3}$$

The total elastic strain energy of the beam at G in fig. 1, which is stored in the elastically deformed core and the inelastically deformed microcracked layers, follows from fig. 2 with equations (1-3) as

$$U_G = B \int_0^a \left[\int_{-\xi h}^{+\xi h} \frac{y^2 S_m^2}{\xi^2 h^2 2E} + \int_{\xi h}^h 2\frac{S_m^2}{2E_m} \right] dy\, dx \tag{4}$$

With the parameter

$$\zeta = 1 - \frac{\psi}{h} \tag{5}$$

equation (4) is written after integration as

$$U_G = \frac{BS_m^2 ha}{3E} \left(1 + \frac{\psi}{h} \frac{3E - E_m}{E_m}\right) \tag{6}$$

On the contrary, the elastic strain energy during bending, which is area OED in fig. 1, follows with equ. (3) as

$$U_{el} = \frac{M\theta}{2} \tag{7}$$

With the curvature of the loaded beam and the moment of inertia I, given as

$$\frac{1}{\rho} = \frac{\theta}{a} = \frac{M}{EI} \quad = \quad \frac{3}{2}\frac{M}{EBh^2} \tag{8}$$

it follows for the strain energy of the beam

$$U_{el} = \frac{BS_m^2 ha}{3E}\left(1 + 2\frac{\psi}{h}\right) \tag{9}$$

thereby neglecting terms of high order. After unloading from G, the recovered elastic energy is given by the work area OGF. This is smaller than the elastic energy, which is stored at G by the amount of residual elastic strain energy, which is still contained in the unloaded beam at F, and is given as

$$U_r = U_G - U_{el} = \frac{BS_m^2 a}{E}\psi\, \frac{E - E_m}{E_m} \tag{10}$$

During microcracking, energy is dissipated in the dilated process zone. The work done per unit volume is $\int S_m d\epsilon$. In the inelastic region $|y| > \zeta h$, the total strain is

$$\epsilon = \frac{y}{\rho} = \frac{S_m y}{E_m \xi h} \tag{11}$$

The inelastic strain is the total strain, less the elastic strain at $|y| = \zeta h$, which is $\frac{S_m}{E_m}$, and for the dissipated energy we obtain

$$U_{diss} = \int W_{diss} dV_{diss} = 2B \int_0^a \int_{yh}^h S_m \left(\frac{S_m y}{E_m \xi h} - \frac{S_m}{E_m}\right) dy\, dx \tag{12}$$

Integration yields

$$U_{diss} = \frac{BS_m^2 a}{hE_m}\psi^2 \tag{13}$$

thereby neglecting terms of higher order.

From the previous section it is obvious, that after unloading of a nonlinear deformed beam, a high amount of elastic strain energy is still stored in the beam. The amount, which depends on the strain energy difference between the elastic damaged process zone and the surrounding safe material at the constant cohesion strength S_m, which is the ultimate uniaxial tensile strength of the material, and arises in the volume of the process zone during incremental crack growth. Thus, the experimental and theoretical results clearly show that damage and failure of nonlinear elastic materials are associated with residual stresses of the first kind. Therefore, a concept for the constitutive description of deformation, including damage and failure, cannot contain a yield condition modified to account for the microcrack porosity. This contradicts the evaluation of zones of residual elastic strains during deformation, and the character of the ultimate tension strength S_m of the material. Failure of nonlinear elastic materials seems not to be dependent on a global energy criterion. Experimental results of damage, and failure of nonlinear elastic materials show the local character of such events. As will be shown in the next section, the crack stability of

232

nonlinear elastic materials is governed by the capability of the material to store a high amount of residual elastic strain energy.

2.2 The microcrack induced residual strain energy release rate as the effective crack driving force.

For otherwise identical beams with different length, which are the starter crack length, the OYGK loading traces are different on a MΘ-plot (fig. 1) because of the different starting stiffness. In terms of a J_c-criterion for initiation of instability, the areas under MΘ curves at various G would be plotted against the crack area, and J determined from the slopes. J_c would be the value of J, at which fracture initiates, since the cracking moments at G are known, and they are independent of crack length. However, the rotation at which they occur depends on the starter crack length a. For plastically damaged materials this was exemplified in detail by Atkins and Mai [8]. For elastically damaged materials, the total elastic strain energy, which is stored in the specimen at crack initiation, follows equ. (9-12) as

$$U_G = U_g + U_{diss} = U_{el} + U_r + U_{diss}$$

$$= \frac{BS_m^2 a}{3E} \psi \left(\frac{h}{\psi} + \frac{3E - E_m}{E_m} + \frac{E}{E_m} \frac{\psi}{h} \right)$$

(14)

This equation shows that the total elastic strain energy U_G mainly depends on two terms. In addition to the size of the damage zone ψ, one is related to the more global geometric parameters, whereas the other depends on the relative change of Young's modulus due to damage.

From theoretical evaluations, given by Kachanov et al. and others [9] about the influence of microcracks surrounding a macroscopic crack tip on the stress field, it can be concluded that only the very near microcracks at the crack front influences the SIF. Therefore, the size of the total damaged volume has a minor influence, and only the density and the elastic interaction of the nearest microcracks is important. This can also be concluded from experimental measurements about the distribution of microcracks in the fracture zone of an alumina specimen by small angle X-ray scattering (SAXS) by Babilon et al. [10]. As can be seen from first results, the size of the damage zone does not vary remarkably with the rising crack resistance during stable crack growth. Therefore, the elastic crack driving force of an elastically damaged material can be expressed by a local criterion, given by the residual strain energy release rate in the following form

$$J_{res} = \frac{S_m^2}{E} \frac{E - E_m}{E_m} \psi$$

(15)

thereby neglecting terms of higher order in ψ/h. With equ. (15) the instability criterion is given as

$$\frac{dJ_{res}}{da} = S_m(\epsilon_m - \epsilon)$$

(16)

therefore, fracture occurs at a critical CTOD of the damage zone [2]. Equ. (15) also follows from the path independent J- integral in the formulation given by Bui and Ehrlacher [11], which was explained in previous works [1,2].

3   Crack resistance at crack initiation

In previous papers it was shown [1,2] that the crack resistance of elastically damaged materials can be expressed in terms of the total elastic strain energy of all microcracks with a mean length of $2a_m$, which are generated in the damage zone during an incremental crack growth of Δa. The elastic strain energy density of one microcrack is $2\gamma/2a_m$. Then, with the volume density of cracked facets β, the total released elastic strain energy in the volume of the process zone during an incremental crack growth is given as

$$U_m = B \int_0^{\Delta a} \int_{-\psi}^{+\psi} \frac{2\gamma}{2a_m} \, dy \, dx \qquad (17)$$

and after integration we get

$$U_m = \frac{2\gamma}{a_m} B\psi \Delta a \qquad (18)$$

The microcrack induced elastic energy release rate equals the change of the residual elastic strain energy (equ. (15)) in the volume element $B2\psi\Delta a$. With equ. (15-17) we obtain a formula for the cohesion strength of the damaged zone in the following form

$$S_m = \sqrt{\frac{2\gamma E}{a_m}} \; I_m \qquad (19)$$

which exhibits a Griffith term, and in addition with the second one, it characterizes the contribution of the change of the residual elastic energy to crack growth, given as

$$I_m = \sqrt{\beta \frac{E_m}{E - E_m}} \qquad (20)$$

Therefore, the cohesion strength increases with decreasing grain facet size, with the microcrack density of a specific microcrack field, and a limited decrease of the elastic modulus. The term I_m was introduced as the elastic interaction parameter, which includes the elastic interaction between the microcracks and the macrocracks to the crack resistance [1].

Physically I_m characterizes the ratio of the SIF's of stress fields without and with microcracks K_0 and K_m, respectively, given as

$$I_m = \frac{K_o}{K_m} \qquad (21)$$

Then, for the fracture energy release rate in materials with elastic damage at crack initiation, the Neuber/Irwin relation is written as

$$\tilde{G}_o = \frac{S_m^2}{E} \; \psi \qquad (22)$$

4 Crack resistance during stable crack growth

Following experimental results, the stress tensor on the surface of the damage zone rotates up to 90^0 at instability due to unloading at a critical residual strain energy of the process zone. Thus, the incremental change of this quantity follows as

$$\tilde{G}_{res} = \psi \, \frac{d}{da} \, \frac{S_m^2}{E} \, \psi \tag{23}$$

and the crack resistance during stable crack growth is the sum of an elastic and an inelastic term with equ. (22,23) as

$$\tilde{G}_R = \frac{S_m^2}{E} \, \psi \, (1 + \frac{E}{S_m^2} \, \frac{d}{da} \, \frac{S_m^2}{E} \psi \} \tag{24}$$

This equation is equivalent to the following form

$$J_R = J(1 + \frac{E}{S_m^2} \, T_J) \tag{25}$$

which was introduced by Paris et al. [12] with the tearing modulus

$$T_J = \frac{d}{da} J \tag{26}$$

Equation (24) opens the possibility to numerically quantify the crack resistance of ceramics on the basis of a quantitative microstructural analysis, which can be introduced with the cohesion strength such as grain size distribution, Gibb's free surface energy, and residual strain energy due to grain boundary mismatch. First results will be worked out in the next section.

5 Discussion

5.1 General remark

It is noticed that the previous analysis for the crack resistance behavior of brittle materials concerns the apparent fracture toughness, due to the external work. In reality, the residual strain energy, which is stored in the damaged fracture zone, generates an internal bending moment $P_i a = M_i$, which contributes to the fracture work. P_i is negative, as a consequence of the radial compressive stresses inside the process zone. An external load $P = -P_i$ must be used to completely flatten a beam. As mentioned in the literature [8], the internal work $1/2 \, P_i u_r$ must be added to the external fracture work to get the "real" toughness. It is the internal bending moment which generates the residual offset u_r after unloading. In many cases P_i is not constant during stable crack growth. Therefore, the stress field on the border of the process zone is also not constant. This will be exemplified in detail in a forthcoming paper.

5.2 Numerical equivalence with measured results

With the relations from previous sections, we will numerically argue that residual stresses govern the crack resistance of elastically damaged materials. Thus, using physical and microstructural data of a material, we can compare measured macroscopic with microscopic parameters, which are listed in the table.

For a first approximation we assume a value of the Gibb's free surface energy $\gamma = 1$ N/m and for $I_m \sim 1$, which is the asymptotic value if β becomes zero and $E_m = E$. With a mean grain size of 10 μm and a size of the process zone of 100 μm [4,5,10] for a specific alumina, we get $G_o = 20$ N/m for the elastic fracture energy release rate at crack initiation (equ. (22)). The cohesion strength follows as $S_m = 270$ MPa. These values are comparable with measured values of similar materials [1,4,5].

To estimate J_{res} and I_m to a first approximation, a value of $E_m = 0,9$ E [13] will be used. Then, the residual strain energy release rate (equ. (15)) which, as a consequence, is due to local unloading of a quantized crack step, and follows as $J_{res} = 2,2$ N/m. This value is in the range of the Gibb's free surface energy and confirms the local character of crack instability. It is noticed by Atkins and Mai [8] that, although small, the residual strain energy has a pronounced influence on the failure of plastically damaged materials.

With the elastic fracture energy release rate at initiation G_o, and the residual fracture energy release rate during stable crack growth, the crack resistance can be written as

$$\tilde{G}_R = \tilde{G}_o + \tilde{G}_{res} \tag{27}$$

Thus, with equ. (24) macroscopic and microscopic features can be compared with each other as [3,6]

$$\tilde{G}_o = \frac{P^2}{2B}\frac{dC_s}{da}$$
$$= \frac{S_m^2}{E}\,\psi = \frac{2\gamma}{a_m}\,\psi\,I_m^2 \tag{28}$$

and

$$\tilde{G}_{res} = \frac{P}{B}\frac{du_r}{da}$$
$$= \psi\,\frac{d}{da}\left(\frac{S_m^2}{E}\,\psi\right) = \psi\,\frac{d}{da}\left(\frac{2\gamma}{a_m}\,\psi\,I_m^2\right) \tag{29}$$

Most important for the toughness of high tech ceramics, is the rate of the stored residual energy $\tilde{G}_{res}$, which shows up to the maximum load, a nearly linear dependency of crack growth Δa. Then, with measured values of $\tilde{G}_{res}$, the total stored residual elastic strain energy, after a stable crack growth of Δa, follows as

$$U_{res} = \int_o^{\Delta a} \tilde{G}_{res}\,da$$
$$= \int_o^{max} \frac{P}{B}\,du_r \tag{30}$$
$$= \psi\int_o^{max} d\left(\frac{2\gamma}{a_m}\,\psi\,I_m^2\right)$$

Table: Material properties of alumina

$2a_m$ mean grain size [4,10], 2β Gibbs free surface energy, 2ψ size of process zone [4,10], E. Youngs' modulus [4,10] $\tilde{G}_0$ elastic fracture energy release rate at crack initiation (equ. (22)), $\tilde{G}_{res}$ residual fracture energy release rate during stable crack growth (equ. (29)), S_m cohesion strength (equ. (19)), $\tilde{G}_{res\ max}$ maximal value of the residual fracture energy release rate of a specific alumina before and after thermal cycling [5].

crack initiation

crack propagation

$2\,a_m$ μm	$2\,\gamma$ N/m	E GPa	S_m MPa	I_m
20	2	365	270	1

$\tilde{G}_{res\ max}$ N/m	I_m
100	5,1
75	4,7

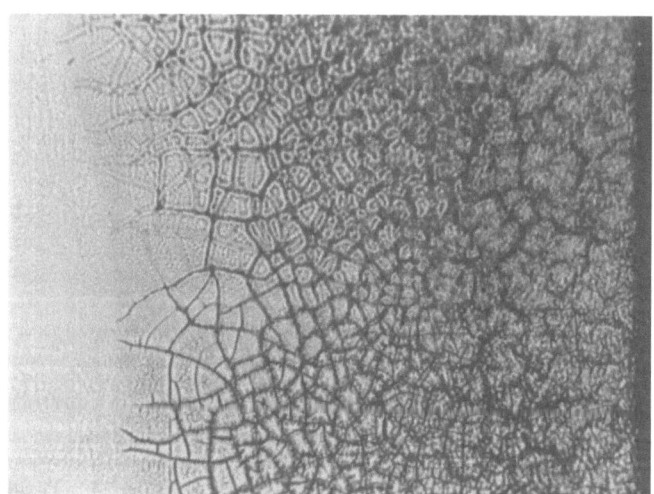

Figure 3: Microcrack pattern in the fracture surface of a glass ceramic [14].

From this we can evaluate a value of I_m, which characterizes the crack shielding effect of the microcrack field. As a first approximation, we will assume constant values of γ, a_m and ψ.

Tertel and Osterstock et al. [5] measured the crack resistance of the same alumina. After stable crack growth of about $\Delta a = 1,5$ mm, a maximum value for $\tilde{G}_{R\ max}$ in the range of 130 N/m was estimated. The respective value for G_{res} was found to be in the range $\tilde{G}_{res\ max} = 100$ N/m. However, this value decreases to 75 N/m after thermal cycling, due to microcracking which was induced by the thermomechanical stresses. With these measured values, we can estimate the respective values of I_m to be 5.1 and 4.7, respectively using equ. (30) (see table). These values seem to be high in comparison to a value of $I_m = 1.6$, which we can estimate, if we assume values for the microcrack density and the reduced Young's modulus $\beta = 0.3$ and $E_m = 0.9$ E, respectively. Only with very high microcrack densities in the range of 0.5 and a reduced Young's modulus of 0.98, a value for I_m, in the range of 5 can be estimated. Thus, further work must be done to clarify elastic damage in brittle materials. Besides the optimization of the analysis, this concerns the numerical and experimental characterization of the effective microcrack configuration, which governs the instability of the process zone, with respect to the stress distribution of the elastically damaged beam.

As mentioned by Kachanov et al. [9], mode II microcracks have a pronounced influence on the SIF of the external stress field. Fig. 3 shows such a mode II microcracks in the fracture surface of a bend specimen of a glass ceramic [14]. From the hydrostatic point in the middle of the fracture surface to the specimen boundaries, the microcrack density increases with decreasing crack velocity. No microcracks can be seen in the middle of the specimen. In addition, the experimental results clearly show evidence of an increasing R- curve for this material, which seems to be attributed to the elastic interaction effect of the observed mode II microcracks with the macrocrack. Contrary to, as shown by Saxs [10], microcracks in alumina are mainly oriented parallel to the ligament. However, by using the decoration technique [14], mode II microcracks can also be seen. Thus, it seems obvious from the analysis that elastic damage induces residual stresses, which governs the crack resistance of brittle materials.

The author would like to thank the DFG for the financial support.

References

[1] F.E. Buresch; Materialprüfung 29, 1987, 261-268

[2] F.E. Buresch; Euro-Ceramics, Vol. 3, eds. G. De With, R.A. Terpstra, R. Metselaar, Elsevier, 1989, 3, 136

[3] F.E. Buresch, K. Frye and Th. Müller; Fracture Mechanics of Ceramics 5, Edited by R.C. Bradt, A.G. Evans, D.P.H. Hasselmann, and F.F. Lange (Plenum Publishing Corporation, 1983) 591-608

[4] K.K.O. Bär, R. Mergen, F. Osterstock; Euro-Ceramics, Vol 3, eds. G. de With, R.A. Terpstra, R. Metselaar, Elsevier, 1989, 3, 190

[5] A. Tertel, 2. Studienarbeit, Clausthal, Nov. 1987; F. Osterstock, R. Moussa; Fortschr.Ber. der DKG, Bd. 3, 1988, 71, Heft 3, ISSN0177-6983

[6] F.E. Buresch, E. Babilon, G. Kleist; Elsevier Applied Science, London 1989, 1003

[7] M. Rühle, A.G. Evans, R.M. McKeeking, P.G. Charalambides, J.W. Hutchinson; Acta metall. Vol. 35, No. 11, pp. 2701-2710

[8] A.G. Atkins, Y.W. Mai; Ellis Horwood Ital, Chichester, 1985

[9] E. Montgaut, M. Kachanov; International Journal of Fracture: R55-R62 (1988)

[10] E. Babilon, K.K.O. Bär, G. Kleist, H. Nickel; Euro-Ceramics, Vol 3, eds. G. de With, R.A. Terpstra, R. Metselaar, Elsevier, 1989, 3, 247

[11] H.M. Bui, A. Ehrlacher; Advance in Fracture Research, ICF 5, Vol. 2, 1981, 533-551

[12] P.C. Paris, H. Tada, A. Zahoor, H. Ernst; American Society for Testing and Materials, 1979, pp. 5-36

[13] A. Quinten, W. Arnold; Conf.Europ.Mat.Res.Soc. Strassburg 1989, Journ. Mat.Sc.Eng. A 122, 1989, 15-19

[14] O. Buresch, F.E. Buresch, W. Hönle, H.G.v.Schnering; Mikrochim. Acta [Wien] 1987, I, 219-224

Stress concentration in microinhomogeneous inelastic materials

P. LIPINSKI, M. BERVEILLER

Laboratoire de Physique et Mécanique des Matériaux (UA CNRS)
ISGMP - ENIM
Ile du Saulcy 57045 Metz Cedex

Summary

This paper deals with the modelling of the inelastic thermomechanical behavior of micro inhomogeneous materials within the framework of the classical uncoupled thermo-elasto-plasticity without damage. At first, the local behavior of material is presented, followed by the global or overall description. A general approach to the determination of concentration tensors from the integral equation is proposed and next specialized for the case of the self-consistent scheme. A few applications are presented concerning elasticity (evaluation of the overall elasticity tensor), thermoelasticity (local stresses, thermal expansion coefficients,...) and elastoplasticity (stress-strain curves, residual stresses...).

Introduction

The evaluation of the thermo-elastic (linear) behavior of microinhomogeneous materials such as polycrystals or composites was studied by many authors, for example Levin [1], Hill [2], Kröner [3]. The theoretical solution of this type of problems is known. The question becomes different when dealing with inelastic microinhomogeneous materials. At first, plastic strain is a supplementary source of internal stresses and at the same time, the ductility of parts of the medium can reduce internal stresses by plastic accommodation.

The aim of this study is to evaluate the local stress fields in the microinhomogeneous solid and to find the thermomechanical overall behavior of such kind of solids. The Medium is considered as continuum, microinhomogeneous and macrohomogeneous. The uncoupled thermo-elasto-plastic behavior is supposed and damage is not taken into consideration. The mechanism of crystallographic slip is adopted in order to describe the plastic deformation of the matrix.

The general form of local relations is reviewed in the first section. In this approach, the temperature acts by the intermediary of thermal expansion coefficients. Introducing the concentration tensors linking local fields with the overall ones and applying the usual averaging operations, the overall behavior may be determined.

In the next section, an integral equation is presented allowing the determination of the concentration tensors. The self-consistent approximation of these tensors is deduced from the integral equation. Finally, theoretical and numerical results concerning the behavior of Metal Matrix Composites are presented. The overall elastic moduli and thermal expansion coefficients as well as concentration tensors are determined for Al-SiC composites. Elastoplastic behavior and residual stresses are discussed for the same composite.

Local constitutive equations

The infinite microinhomogeneous medium of volume V undergoes a thermomechanical loading defined by the state of overall stresses Σ, the temperature θ and the corresponding rates $\dot{\Sigma}$ and $\dot{\theta}$.

At a point r of the medium, various physical mechanisms contribute to the total strain rate $\dot{\varepsilon}^T(r)$ i.e. :

* The elastic deformation defined by the local stiffness s(r) and stress rates $\sigma(r)$

$$\dot{\varepsilon}^e_{ij}(r) = s_{ijkl}(r)\,\dot{\sigma}_{kl}(r) \tag{1}$$

* The thermal deformation described by local thermal expansion coefficients $\alpha_{ij}(r)$ and temperature rate $\dot{\theta}$ assumed to be homogeneous through the considered volume V

$$\dot{\varepsilon}^{th}_{ij}(r) = \alpha_{ij}(r)\,\dot{\theta} \tag{2}$$

* The plastic deformation modelled by the local plastic stiffness tensor P(r) submitted to the local yield conditions

$$\dot{\varepsilon}^P_{ij}(r) = P_{ijkl}(r)\,\dot{\sigma}_{kl}(r)$$
$$f(\sigma_{ij}(r)) = 0 \quad \text{and} \quad \frac{\partial f}{\partial \sigma_{ij}}\,d\sigma_{ij} > 0 \tag{3}$$

The total deformation rate $\dot{\varepsilon}_{ij}^T(r)$ is then :

$$\dot{\varepsilon}^T_{ij}(r) = \dot{\varepsilon}^e_{ij}(r) + \dot{\varepsilon}^{th}_{ij}(r) + \dot{\varepsilon}^P_{ij}(r) \tag{4}$$

A more general case of a constitutive equation including phase transformation was given by Patoor et al [4] and the form of P and f in the case of crystallographic slip can be found in [5].

Introducing (1), (2) and (3) into (4), one has :

$$\dot{\varepsilon}^T_{ij}(r) = g_{ijkl}(r)\,\dot{\sigma}_{kl}(r) + n_{ij}(r)\,\dot{\theta} \tag{5}$$

where

$$g_{ijkl} = s_{ijkl} + P_{ijkl}$$
$$n_{ij} = \alpha_{ij} \tag{6}$$

Global thermomechanical constitutive equations

The global thermomechanical constitutive law takes a form analogous to (5)

$$\dot{E}^T_{ij} = G_{ijkl}\,\dot{\Sigma}_{kl} + N_{ij}\,\dot{\theta} \tag{7}$$

Where the overall stress and strain rate $\dot{\Sigma}$ and $\dot{E}^T$ may be deduced from the local ones by the usual averaging operations

$$\dot{\Sigma}_{ij} = \frac{1}{V} \int_V \dot{\sigma}_{ij} (r) \, dV \equiv \overline{\dot{\sigma}}_{ij} (r) \tag{8}$$

$$\dot{E}^T_{ij} = \frac{1}{V} \int_V \dot{\varepsilon}^T_{ij} (r) \, dV \equiv \overline{\dot{\varepsilon}}_{ij} (r) \tag{9}$$

The dual form of equations (5) and (7) may be written as

$$\dot{\sigma}_{ij} (r) = l_{ijkl} (r) \, \dot{\varepsilon}^T_{kl} (r) - m_{ij} (r) \, \dot{\theta} \tag{10}$$

$$\dot{\Sigma}_{ij} = L_{ijkl} \, \dot{E}^T_{kl} - M_{ij} \, \dot{\theta} \tag{11}$$

where

$$\begin{array}{ll} l = g^{-1} & m = l : n \\ L = G^{-1} & M = L : N \end{array} \qquad \begin{array}{l} (12) \\ (13) \end{array}$$

In order to obtain the overall behavior of the material using the local constitutive relations and knowing the microstructure of the composite, two concentration relations are introduced

$$\dot{\sigma}_{ij} (r) = B_{ijkl} (r) \, \dot{\Sigma}_{kl} + b_{ij} (r) \, \dot{\theta}$$

$$\dot{\varepsilon}^T_{ij} (r) = A_{ijkl} (r) \, \dot{E}^T_{kl} + a_{ij} (r) \, \dot{\theta} \tag{14}$$

To obtain the overall thermomechanical behavior i.e. L and M tensors, one has

$$\dot{\sigma}_{ij} (r) = l_{ijkl} (r) \, [A_{klmn} (r) \, \dot{E}^T_{mn} + a_{kl} (r) \, \dot{\theta}] - m_{ij} (r) \, \dot{\theta} \tag{15}$$

Applying the averaging operations

$$\dot{\Sigma}_{ij} = \overline{l_{ijkl} (r) \, A_{klmn} (r)} \, \dot{E}^T_{mn} + \overline{l_{ijkl} (r) \, a_{kl} (r) - m_{ij} (r)} \, \dot{\theta} \tag{16}$$

it follows

$$L_{ijmn} = \overline{l_{ijkl} (r) \, A_{klmn} (r)}$$

$$M_{ij} = \overline{l_{ijkl} (r) \, a_{kl} (r) - m_{ij} (r)} \tag{17}$$

Similar calculations allow to determine G and N in function of g, n, B and b.

In order to find the concentration tensors A and a, several methods may be proposed. Nevertheless, they constitute only an approximation of the integral equation proposed in the following section.

Determination of Concentration Tensors

The general approach is to consider the inhomogeneous solid as a continum medium with a microstructure satisfying

- The equilibrium equation

$$\dot{\sigma}_{ij,j} (r) = 0 \tag{18}$$

- The constitutive equation

$$\dot{\sigma}_{ij} (r) = l_{ijkl} (r) \, \dot{\epsilon}^{T}_{kl} (r) - m_{ij} (r) \, \dot{\theta} \qquad (19)$$

- The compatibility equation

$$\dot{\epsilon}^{T}_{ij} (r) = \frac{1}{2} (\dot{u}^{T}_{i,j} + \dot{u}^{T}_{j,i}) \qquad (20)$$

Introducing a homogeneous medium (without a microstructure) and characterized by its tangent moduli tensor L° such that

$$l (r) = L^\circ + \delta l (r)$$

equations (18) (19) and (20) take the form

$$L^{\circ}_{ijkl} \, \dot{u}^{T}_{k,lj} + [\delta l_{ijkl} (r) \, \dot{\epsilon}^{T}_{kl} (r) - m_{ij} (r) \, \dot{\theta}]_{,j} = 0 \qquad (21)$$

Using the Green tensor technique, (21) may be transformed into an integral equation [3]

$$\dot{\epsilon}^{T}_{ij}(r) = \dot{\epsilon}^{\circ}_{ij} + \int_{V} \Gamma_{ijkl} (r\text{-}r') \, [\delta l_{klmn}(r') \, \dot{\epsilon}^{T}_{mn}(r') - m_{kl}(r') \, \dot{\theta}] \, dV' \qquad (22)$$

where
- $\dot{\epsilon}^{\circ}_{ij}$ is the strain rate of the homogeneous medium submitted to the same boundary conditions as the effective medium
- $\Gamma_{mnij} = 1/2 \, (G_{mi,jn} + G_{ni,jm})$
- G_{mi} is the Green tensor of the infinite medium with L° moduli tensor.

Relation (22) constitutes an integral equation linking the local strain rate $\dot{\epsilon}^{T}$ and the thermomechanical loading characterized by $\dot{\epsilon}^{\circ}$ and $\dot{\theta}$. To exploit this complex equation, various approximations have to be applied leading to numerous models. In this work, the self-consistent approach is developed and applied. This method may be shortly characterized as follows :

* the interactions between constituents, in the case of one-site scheme, are modelled by interactions between an ellipsoïdal inclusion and the homogeneous equivalent medium L^{eff} considered as matrix
* the particular choice of $L^\circ = L^{eff}$ is adopted.

The multiple-site version of this approach was established by Berveiller et al [6] and for elastoplastic metals at large strains, the model was generalized by Lipinski and Berveiller [7].

Let us now consider a granular medium and suppose that l, ϵ^{T} and m are piece-wise constant. The following holds

$$\delta l (r) = \sum_{I=1}^{N} \delta l^{I} \, y^{I} (r)$$

$$m (r) = \sum_{I=1}^{N} m^{I} \, y^{I} (r) \qquad (23)$$

$$\dot{\varepsilon}^{T}(r) = \sum_{I=1}^{N} \dot{\varepsilon}^{TI} y^{I}(r)$$

where $y^I(r)$ represent N characteristic functions such that

$$y^I(r) = \begin{cases} 0 & \text{if } r \notin V^I \\ 1 & \text{if } r \in V^I \end{cases} \tag{24}$$

In order to simplify the considerations, suppose for the moment that $\dot{\theta} = 0$. In this case, the integral equation is rewritten in the form

$$\dot{\varepsilon}^{TI}_{ij} = \dot{\varepsilon}^{o}_{ij} + T^{II}_{ijkl} \delta l^{I}_{klmn} \dot{\varepsilon}^{TI}_{mn} \tag{25}$$

where

$$T^{II}_{ijkl} = \frac{1}{V_I} \int_{V_I} \int_{V_I} \Gamma_{ijkl}(r-r') \, dVdV' \tag{26}$$

By definition, $L^o = L^{eff}$ which leads to

$$\dot{\varepsilon}^o = \dot{E}^T \tag{27}$$

and the concentration tensor A^I for the Ith inclusion is

$$A^{I}_{mnkl} = (I_{mnkl} - T^{II}_{mnij} \delta l^{I}_{ijkl})^{-1} \tag{28}$$

The tangent moduli tensor L^{eff} may be obtained from (17) and (28)

$$L^{eff}_{ijkl} = \sum_{I=1}^{N} F^I l^{I}_{ijmn} A^{I}_{mnkl} \tag{29}$$

where F^I denotes the volumic fraction of the considered phase. In the similar manner, one may determine a^I and using (17)

$$M^{eff}_{ij} = \Sigma \, F^I (l^{I}_{ijkl} a^{I}_{kl} - m^{I}_{ij}) \tag{30}$$

Expressions (29) and (30) are rather complex ones because of their implicit character.
The proposed theory has been applied to elasticity, thermoelasticity and elastoplasticity of Al-SiC metal matrix composites.

Application to Al-SiC composites

The composite is considered as a two-phased material composed of a matrix and reinforcements. The first application corresponds to an experimental study by Schneider et al [8]. The properties of the matrix and reinforcements have been deduced from this study [8] and are presented in Table 1.

The material was extruded and exhibits a small anisotropy. The reinforcements presented an ellipsoïdal aspect elongated in the extrusion direction. The numerical calculations of Young's modulus in the transversal direction using the self-consistent model are shown in Fig 1. The results agree very well with the experimental measurements. An accurate modelling of the elastic anisotropy has been stated.

	Young's Modulus E	Poissons's ratio v
Al	77.5	0.3
SiC	430	0.3

Table 1 : Elastic properties of Al-Matrix and SiC reinforcements.

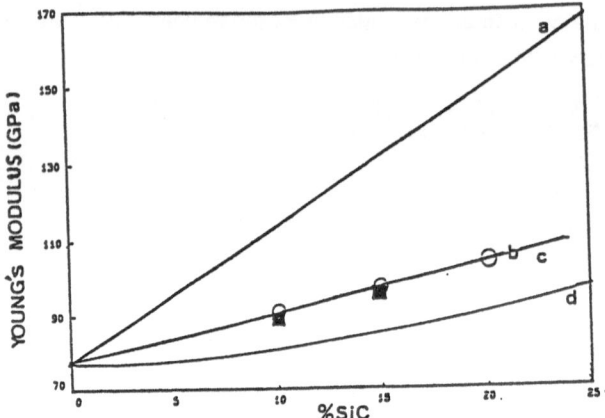

Figure 1 : Comparison of the transversal Young's modulus : (O) calculated by the self-consistent model, (■) experimentaly obtained by Schneider et al [8] and determined using (a) Voigt and (b) Reuss models

Next, the thermoelastic behavior of Al-SiC composites the local thermal expansion tensors have been supposed isotropic both for matrix and SiC (α_{Al} = 21 10^{-6} K^{-1}, α_{SiC} = 4 10^{-6} K^{-1}). Fig 2 presents the evolution of α_{33}^{eff} component of the effective ther - mal expansion tensor for four reinforcement shapes and various F^{I}.

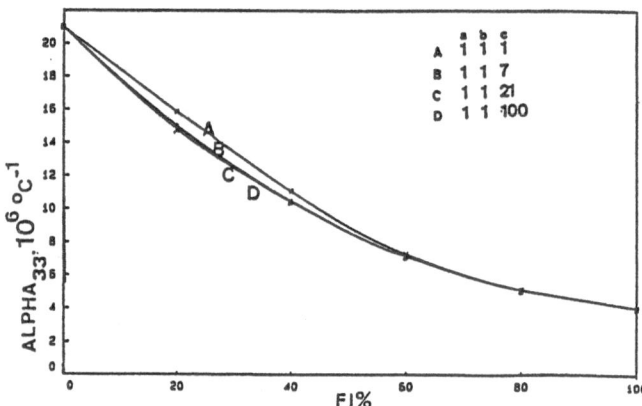

Figure 2 : Evolution of the α_{33}^{eff} component of the effective thermal expansion tensor as a function of the volumic fraction

The components of the concentration tensors B_{3333} and b_{33} are plotted in Fig 3 and 4. A very strong stress concentration is observed in both cases depending on the volume fraction and reinforcements shape.

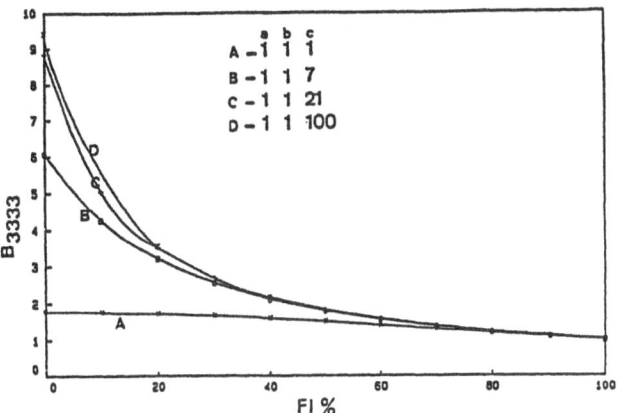

Figure 3 :Evolution of B_{3333} component of the concentration tensor as a function of the volumic fraction and various fiber aspects

The last class of applications concerns the elastoplastic behavior of Al-SiC MMC. The material properties are summarized in Table II.

In this case, the composite is considered as a microinhomogeneous medium with two constituents

- the elastic and isotropic reinforcements
- the microinhomogeneous polycrystalline matrix composed of grains crystallo graphically misoriented.

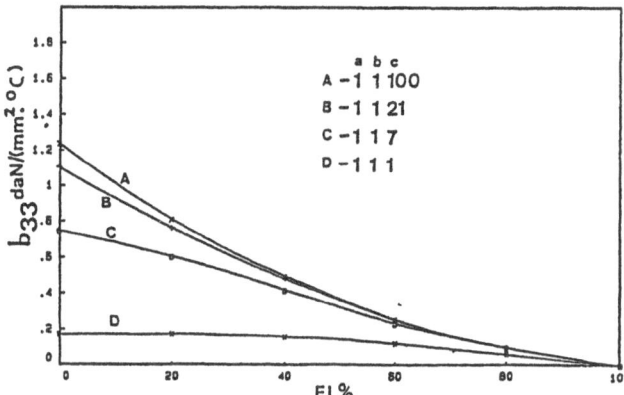

Figure 4 :Evolution of b_{33} component of the concentration tensor as a function of the volumic fraction and various fiber aspects

		SiC	Al
Elasticity	μ ν	$\mu = 27559$ daN/mm^2 0.27	2692 daN/mm^2 0.3
Plasticity	τ_o	∞	14 daN/mm^2
	slip systems Hardening matrix	- -	(111) < 110 > (12x) $H_1 = \mu/250$ $H_2 = H_1 \times 3$
Microstructure	Volumic Fraction	F_I	$(1 - F_I)$
	Form and	elliposoïdal (a,b,c)	spherical
	orientation	same orientation	100 crystallographic orientation (random)

Table II : Mechanical properties and microstructure of the studied composite

The imposed loading corresponds to a tension test for three volume fractions. The influence of the reinforcements contents is very pronounced (Fig 5).

The evolution of residual stresses (σ_{33} components) for some particular grains of the matrix and SiC fiber as a function of the volume fraction of SiC and plastic strain is plotted in Fig 6. The intragranular inhomogeneities appear negligible with respect to the difference of the mechanical behavior between matrix and fiber.

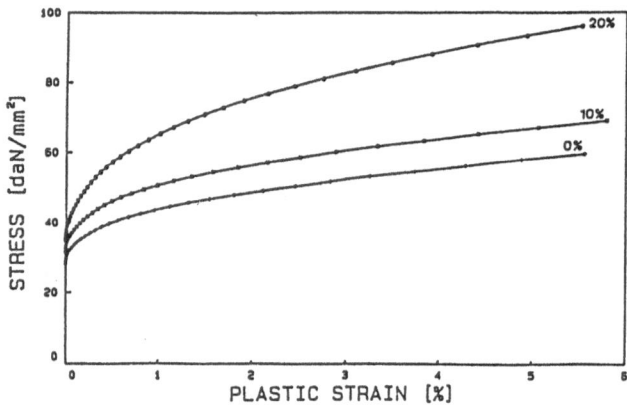

Figure 5 :Tensile curves for 0, 10 and 20% of fibers of spherical shape (a=b=c)

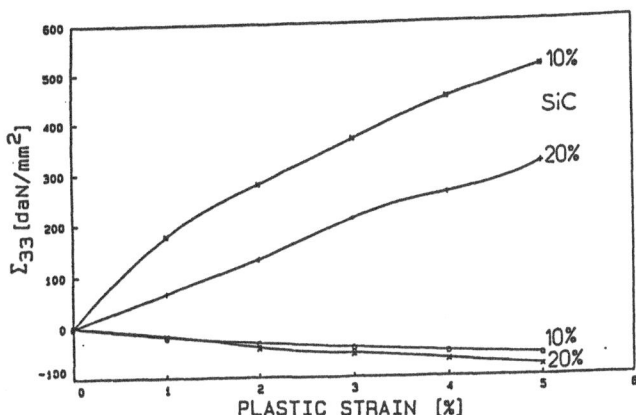

Figure 6 :Residual stress (σ_{33}) inside the reinforcements and particular grains of the matrix as a function of the overall plastic strain and volumic fraction of fibers for ellipsoïdal reinforcements (a = b = c/10)

<u>References</u>

1.   Levin V.M. : Thermal expansion coefficients of heterogeneous materials. Mekhanika Tverd. Tela (1967), 88-94.
2.   Hill R. : Elastic properties of Reinforced Solids. J. Mech. Phys. Solids (1963), 11,

      357-372.
3.   Kröner E. : Zur klassischen Theorie statistisch aufgebauter Festkörper. Int. J. Engng. Sci. (1973), 11, 171-191.
4.   Patoor E., Eberhardt A., Berveiller M. : An integral equation for the polycrystalline thermomechanical behavior of shape memory alloys. Int. Seminar Mecamat Besançon (France), 1988, 319-330.
5.   Beradai C., Lipinski P., Berveiller M. : Plasticity of metallic polycrystals under complex loading path. Int. J. Plasticity (1987), 3, 143-162.
6.   Berveiller M., Hihi A., Fassi Fehri O. : Multiple site self-consistent scheme. Int. J. Engng. Sci. Vol 27 n° 5 p 495-502 (1989).
7.   Lipinski P., Berveiller M. : Elastoplasticity of microinhomogeneous metals at large strains. J. Plasticity. Vol 5 p 149-172 (1989).
8.   Schneider E., Lee D., Razvi S., Salama K. : Non destructive characterization of metal-matrix composites. Proc. 16th Symposium of Non Dest. Evaluation. San Antonio USA (1987).

Lecture Notes in Engineering

Edited by C.A. Brebbia and S.A. Orszag

Lecture Notes in Engineering

Edited by C.A. Brebbia and S.A. Orszag

Vol. 59: K. P. Herrmann, Z. S. Olesiak (Eds.)
Thermal Effects in Fracture
of Multiphase Materials
Proceedings of the Euromech Colloquium 255
October 31 - November 2, 1989, Paderborn, FRG
VII, 247 pages. 1990